how to
CUT HEATING
and
COOLING COSTS

how to CUT HEATING and COOLING COSTS

by Peter Jones

Butterick Publishing

Illustrations by *Ken Rice*

Book Design by *Jos. Trautwein*

Library of Congress Cataloging in Publication Data

Jones, Peter, 1934-
   How to cut heating and cooling costs.

   Includes index.
   1. Dwellings—Energy conservation.
   2. Dwellings—Heating and ventilation—Costs.
   3. Dwellings—Air conditioning—Costs. I. Title.
TJ163.5.D86J66   697   79-15207
ISBN 0-88421-092-8

Copyright © 1979 by Butterick Publishing
                708 Third Avenue
                New York, New York 10017
                First Printing, September 1979
                Second Printing, February 1980

Manufactured and printed in the United States of America, published simultaneously in the USA and Canada

contents

1
some
FACTS
and THEORIES
about
ENERGY

Until not too many years ago, people were still heating and cooling their homes the way it had been done for centuries. They built a fire in their fireplaces or wood burning stoves, and when the house became too warm they opened all the doors and windows. Heating and cooling a home today is considerably more complicated and, unfortunately, more expensive (even the price of firewood has become exorbitant). The reason for this increased complexity is that during the past quarter century we have become increasingly aware of the importance of maintaining a proper temperature in our homes during both hot and cold weather, and also of retaining the proper amount of moisture in the air. Temperature and moisture content, it has been discovered, has a significant effect on our health, as well as our comfort.

For most American homeowners there is an inherent problem when trying to achieve and maintain either the correct moisture content or a comfortable temperature in their residences. The problem comes from the fact that until the last decade contractors paid little or no attention to insulating the houses they built. As a result, the majority of the homes in America are, so far as energy conservation is concerned, not much better than a sieve. So to warm or cool most American homes and then keep them at a comfortable temperature requires considerable amounts of electricity, oil, natural gas or coal. And these days all of those energy-producing fuels demand considerable outlays of money.

Nothing in your home costs more to operate than its heating and cooling systems. Americans annually pour billions of dollars into expensive fuels solely for their personal comfort, and the price of those fuels keeps rising. Unfortunately, no home is so completely airtight that all of the energy its owner purchases is ever entirely used. In fact, tremendous amounts of the energy you buy are literally thrown out every window in your house before you even feel the comfort it is supposed to provide. Many homeowners try to combat their loss of warm or cool air by installing larger-than-necessary furnaces or oversized air conditioners, which demand more money because they use more fuel to create more energy—most of which promptly dissipates just as quickly as before.

The irony of all this wasted fuel and money is that you can reduce home energy losses by as much as 60%, for very few dollars and almost none of your time, with a variety of simple do-it-yourself projects. But a 60% reduction in energy loss is only the beginning. It is possible to eliminate practically all fuel costs by constructing passive solar heating and cooling complexes, which cost relatively little to install and require almost no maintenance (you do have to polish the glass once in awhile). But before you start any of these projects it helps to know something about the nature of hot and cold air, and how it acts.

THE NATURE OF HEAT AND COLD

Air is always on the move. If it is cool it tends to be sluggish, staying close to the ground and requiring considerable force to move it. Consequently, if you use an air conditioner to cool the rooms in your house it must have a fairly powerful motor, which, of course, demands a considerable amount of electricity in order to operate.

By contrast, warm air is lighter than cold and is always racing around in all directions. Because it is lighter it invariably rises above any cold air around it, and it requires less force to move it. For example, if you have a forced air heating system there is a motor-driven blower situated near the furnace which draws air through the duct system. If you only need the blower for heating purposes it can be as small as ⅓ horsepower. But if you also need that blower to distribute cool air during the summer, it will have to be driven by at least a ½-horsepower motor.

Warm air, that is heat, flows in three ways: radiation, convection and conduction. *Radiation* is the emission of heat from a warm object, such as a radiator or fireplace. It is the nature of heat to immediately transfer itself away from any warm object to the cooler air around it. *Convection* occurs as heat comes in contact with cooler air and begins to warm up that air. It warms not only the cooler air, but also any cooler object that it touches. But remember that the heat is moving faster than the cooler air and also rising above it, which creates a current. The warmest air in any room, then, is always up near the ceiling, which is why the floors of your house always feel colder than the air around your head and shoulders. Once the warm air gets to the ceiling, or comes in contact with the cooler outside walls and windows of your house, it conducts itself right outside. *Conduction* occurs when heat leaves any warm object and moves toward anything cooler (less warm). Put a warm and a cold object next to each other and heat from the warm object will immediately move toward the colder one. It is conduction that constantly pulls whatever heat is in your house through the walls and up to the roof, then to the outside cold. It is also conduction that draws hot air inside your home during a warm summer day.

The flow of energy, then, begins with the release of heat (radiation) into the cooler surrounding air (convection); the heat then moves upward through solid walls (conduction) and to the colder air outside your home.

HEATING AND COOLING SYSTEMS

CENTRAL HEATING

Since heat radiates, conducts and convects all the time, the heating and cooling systems used to control the temperature in our homes are all designed around the principles of radiation, conduction and convection. The two major systems used to centrally heat homes today are known as warm-air (or indirect) heating, or hot-water (direct) heating. With the indirect system, air is warmed by a furnace and then fan-forced through air ducts to various rooms in the house. A hot-water system heats water in a boiler attached to the furnace, which is then forced through pipes to each of the radiators in your home. The water may arrive at the radiators in the form of hot water or, if it has been boiled in the furnace, as steam. In either case, the radiator becomes hot and the heat is immediately radiated, convected and conducted into the surrounding area.

The fuel that runs the furnace may be #2-grade crude oil, natural gas, coal or electricity, all of which are becoming more expensive by the day.

AIR CONDITIONING

Whereas your heating system is hard at work all winter trying to heat the air in your house, in summer air conditioning is used to regulate not only the temperature, when it becomes too high, but also the moisture, cleanliness and movement of air. An air-conditioning system can cool warm air and warm cool air; it can add moisture to as well as remove it from the air, and replace stale air with fresh. Most home air conditioners are powered by electricity, which in most parts of America is already an expensive form of fuel.

For details on heating and cooling systems, see chapters 5 and 6.

KEEPING BILLS DOWN

There are a surprising number of things that a homeowner can do to reduce heating and cooling costs, not all of which involve opening the toolbox. To begin with, there are some energy-saving habits that can be developed:

1 • Turn off the radiators or close air ducts in rooms that are not being used, such as a spare guest room.

2 • Lower the thermostat at night or any time the house is vacant. This can be done manually, but you can also install an automatic thermostat which cuts back the heat after a given hour in the evening, then raises it at a preset time in the morning. Generally the nighttime temperature

should be six or seven degrees below the daytime temperature. A reduction of more than that will demand too much fuel to warm up the house again in the morning.

3 • Keep the dampers to fireplaces closed except when burning a fire.

4 • Draw curtains over all windows in the evening and open them during sunlight hours.

5 • Lock all window sashes to cut down on drafts, by pulling the sashes together as tightly as you can.

6 • Install storm windows and doors unless your house is equipped with insulated (double or triple thickness) windows. Keep them latched.

7 • Check the outside of all doors and windows at least once a year for cracks and caulk any that you find.

8 • If you have a forced air heating system, clean the furnace filter at least once a month. Dirty filters inhibit the movement of warm air.

9 • Have the furnace cleaned and adjusted at least once a year to be sure you are getting balanced heat throughout the distribution system.

These are things that should be done each year as a matter of course, no matter what kind of house you live in. Because they are energy-conserving habits that will work toward lowering your heating and cooling costs they are important habits to observe, but they are the merest beginning toward conserving the energy your home loses each year. The real beginning to energy conservation is to insulate your home as fully as you possibly can.

2 insulating CEILINGS and WALLS

BATTS, BLANKETS, LOOSE FILL AND THE R-VALUE

No matter how a house is built, no matter what the quality of workmanship or the kinds of building materials used, the house will not be airtight unless it is fully insulated. If it isn't, or if the insulation is faulty or insufficient, that nice, comfortable, warm air generated by the furnace will immediately drain outside.

If your home was built before 1940, the chances are there was no insulation whatever put into it because there was no national building code that required insulation. If your home dates from 1940 to very recently, it probably has 1½" of insulation in the attic, as required by the Minimum Property Standards (MPS) of the Federal Housing Administration. The warm air you cook up in your basement on cold wintery days stays inside your house just about as efficiently as it does in pre-1940 homes.

The federal goverment, in the early 1970s, updated its insulation requirements for current home construction to at least 6" of insulation in the attic. But that is only sufficient for homes in mild climates. In colder parts of the country, insulation in the attic should be at least 12" thick. And that's just the attic, which is usually the easiest place to insulate.

Actually, nobody figures insulation in inches anymore, because there are a variety of insulating products on the market, some of which will do a fine job of insulating your house without being 6" or 12" thick. So insulation has graduated to a new standard known as the R-value. R-value is the resistance the insulation has to heat passing through it. The higher the R-value, the better the insulation.

It may seem like a waste of time and money to go around stuffing your walls and ceilings with a couple of hundred dollars' worth of insulation, but all government and industry studies that have been made on insulation demonstrate that however much you spend will probably be saved in fuel costs within the space of one year. You will keep on saving at least that much every year, and probably more, since the cost of fuel annually continues to rise. Moreover, the federal government is so convinced of the importance of insulation that it gives home-owners a tax credit of 30% on the first $750 in insulation expenses. In fact, that 30% is better than a normal tax deduction because you can take it off the total amount of tax you owe.

PLUGGING UP THE DIKE YOU LIVE IN

The objective of insulating a home is to wrap the living area in properly installed insulating materials to insure maximum heat retention. In order to "wrap" the entire living area, there should be insulation in all of the exterior walls and between the top floor ceiling and an unfinished attic, or be-tween the attic and the roof if the attic is used as a living space. The floor above an unfinished basement or crawl space should also be insulated, along with any inside walls that are adjacent to an unheated ga-rage.

The kind of insulation and R-value you use is more or less up to you and your local building code, but there are some guidelines to follow. Look at the map shown on this page and locate the zone you live in; the R-values for insulating ceilings, exterior walls and floors in each zone are shown in the table on the following page.

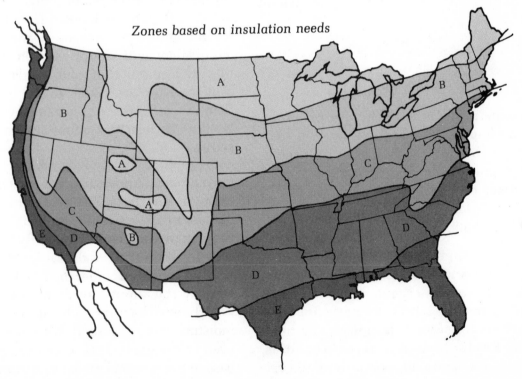

Zones based on insulation needs

Zone	Ceiling	Exterior Wall	Floor
A	R-38	R-19	R-22
B	R-33	R-19	R-22
C	R-30	R-19	R-19
D	R-26	R-19	R-13
E	R-26	R-13	R-11
F	R-19	R-11	R-11

If an insulation, say, fiberglass, has an approximate R-value of 3.3 per inch of thickness and you are living in zone B, you will need to buy insulation that is 10″ thick to meet the minimum requirements for insulating the top floor ceiling of your house. It is to your advantage to put in the highest R-value, thickest insulation you can find, because like everything else, the cost of insulation is going up at the rate of about 10% to 12% a year, which means that in ten years you will pay twice as much as you would today.

INSULATION MATERIALS

The materials sold today for use as insulation in homes are varied not only in their R-values, but also in their ease of installation, degree of nonflammability and numerous other properties. When you are choosing insulation, there are some points to consider about each:

Cellulose•(Approximate R-value: 3.7 per inch of thickness) Cellulose has been long considered a fire hazard, but when it is properly treated, it becomes more fire resistant than rock wool or fiberglass, and has a higher R-value besides. It comes in blankets, batts, and as loose fill. Just be sure the cellulose you purchase has a treatment certification.

Rock wool•(Approximate R-value: 3.3 per inch of thickness) Fiberglass•(Approximate R-value: 3.3 per inch of thickness) About the only disadvantages to fiberglass or rock wool are that they irritate your skin and can develop an odor if they get wet. Otherwise, they are easily installed, inexpensive, and sold just about everywhere in batts, rolls and pellets (for loose-fill installation). They are both fire resistant, but the paper facing in the batts and rolls is usually flammable.

Perlite•(Approximate R-value: 2.7 per inch of thickness) Vermiculite•(Approximate R-value: 2.1 per inch of thickness) Both perlite and vermiculite are inexpensive and easy to install. They are both sold as loose fill that can be poured or blown into place, but they tend to pack down from their own weight after a few years. If they become damp, they will compress even more and thereby lose much of their insulating capabilities. They are better suited in ceilings than in walls, where they can leave uninsulated voids near the ceiling as the material settles.

Polystyrene•(Approximate R-value: 3.5 per inch of thickness) This is sold in rigid panels, which are combustible. It is ideal for use in new construction but only if it is covered by at least ½″ wallboard, exterior siding, or whatever other materials are designated by your local building code. Polystyrene has the advantage of being moisture resistant, and therefore can be used below grade. It is sometimes used as a base under poured-cement slab foundations.

Urea formaldehyde•(Approximate R-value: 4.5 per inch of thickness) *Urethane*• (Approximate R-value: 5.3 per inch of thickness) Urea formaldehyde and urethane have the highest R-values of all the insulators and both are fire resistant, although urethane gives off a noxious gas during a fire. They are both foam-type materials which must be injected into the space they are to insulate, and the equipment needed to inject them is too specialized to be available to homeowners. If they are not installed properly, they will leave a lingering odor that is uncomfortable to live with, so be sure the contractor you hire knows what he is doing.

WHAT GOES WHERE

Batts, which are made of fiberglass or rock wool, are typically 15″ or 23″ wide, to fit between standard joist spacings, and are between 4′ and 8′ long in thicknesses of 1″ to 7″. You can get them with or without a vapor barrier. They are easy to handle, particularly in unfinished attic floors or rafters or basement ceilings.

Blankets are also made of fiberglass and rock wool in the same widths and thicknesses as batts, but they are considerably longer. They can also be used underneath open floors in unfinished attics and basements.

Loose fill can be in the form of pellets or of a much finer consistency made from fiberglass, rock wool, cellulose, vermiculte or perlite. Many of these materials can be poured into an unfinished floor and then raked even. Loose fill is especially useful if there are many obstructions and irregularities in the area being insulated. Loose fill can also be blown into hard-to-reach corners and finished walls and floors, but requires equipment and experience to fill all voids properly. If you intend to blow in loose fill, consider hiring a reputable contractor.

Foam can be used any place in the house and is either urea formaldehyde or urethane. It is expensive and must be installed by a qualified contractor with the proper know-how and equipment.

Rigid boards are always made of polystyrene, urethane or fiberglass. They come in thicknesses of ¾″ to 4″, measure 24″ to 48″ in width, and can be used to insulate a basement wall and then covered with wallboard.

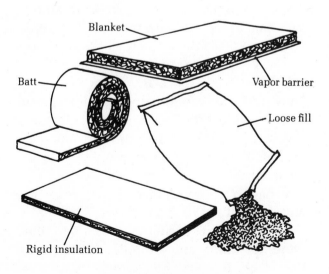

Types of insulation

FIGURING INSULATION NEEDS

It helps considerably if you know where your house is already insulated, and what the insulation material is, so you can estimate its R-value. If you can't figure out what the existing material is, take a sample of it to your insulation dealer.

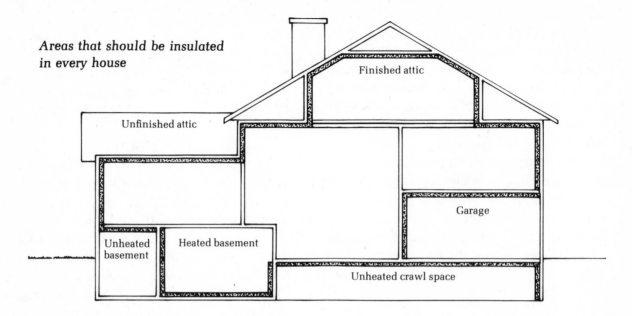

Areas that should be insulated in every house

Attics

Probably the easiest place to investigate insulation is the attic, particularly if it does not have a floor. Attics are of all shapes and contain anything from bare rafters and joists to a completed room. They should have some form of insulation in them somewhere, usually in the floor. If the floor is unfinished, simply poke a ruler into the insulation as far as it will go. If you know what the material is, you can readily estimate its R-value. If you cannot determine the material, take a hunk of it to an insulation dealer. If the R-value is less than it should be for the zone you live in (see map on page 15) you will have to add more insulation.

With a finished or partially finished attic, the insulation could be under the floorboards or in the rafters directly under the roof, the end walls and any dormers that may exist. If you cannot lift one of the floorboards, drill a ½″ or ¾″ hole in an out-of-the-way corner of the floor and peer into it with a flashlight. If the insulation comes up to the floorboards, all you need to do is hook a piece of it with a bent wire to determine what material it is and estimate its R-value.

If the insulation does not come up to the floorboards, poke a wire down to the top of the insulation and mark the wire, then pull it out and measure it. Subtract the thickness of the flooring (¾″ to 1″) and that will tell you how much space is left in the cavity. Now push the wire down through the insulation until it hits the underside of the ceiling, below the floor. Mark the wire and measure it, then subtract the amount of space between the floor and the insulation. When you have retrieved some of the insulation and determined its R-value, you may find it is inadequate. Your alternatives are to blow more insulation under the floor or to rip up the floorboards and add more batts or blankets between the joists.

Walls

To check the insulation in the walls in a finished attic, it may be necessary to poke a hole through an obscure corner to see if there is any insulation behind it. Before you do that, however, turn off the electricity and remove any electrical switch or outlet plate, then poke around the sides of the electrical box. You may have to enlarge a crack along the side of the box with a utility knife, but be careful not to make the hole any wider than can be covered with the plate. When your hole is large enough, shine a flashlight into it. Unfortunately, electricians sometimes push the insulation well away from electrical boxes when they install an outlet, so just because you cannot see any insulation does not mean that it is not there.

You must check each of the dormers, the side walls, end walls and roof rafters in a finished attic before you can be sure of what areas have or have not been insulated.

Most homes are framed with wood studs

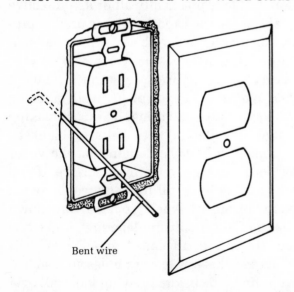

Bent wire

Probe the sides of electrical boxes for insulation

spaced either 16″ or 24″ apart. One side of the studs supports the outside sheathing, a water barrier and the exterior siding, which may be masonry, aluminum, wood or asbestos. The inside of the studs forms a support for lath and plaster or wallboard. Insulation should be stapled between every pair of studs with its moisture barrier facing the heated side of the wall.

In many instances, the simplest method of determining whether or not an outside wall is insulated is to feel it. It should be about room temperature to the touch. If it is cold, you can assume it is uninsulated. You may also be able to feel a draft coming through the electrical outlets and switches if there is no insulation. if you still are not sure, shut off the electricity and remove the faceplates over several of the outlets and switches, then probe around the electrical boxes. If all else fails, drill a small hole into the wall somewhere between the studs and try to hook out some insulation with a bent wire. You can fill in the hole with wallboard compound and tape after you are done.

Basements and Crawl Spaces

Whether your house has a full basement, a crawl space, or was built on a concrete slab, you can reduce your heating and cooling costs by a considerable amount if you insulate the underside of the ground floor. If the basement is unfinished, take a good look at it. If all you can see are bare masonry walls and unadorned wood between the ceiling joists, there is no insulation. In an unheated basement, you do not need insulation on the walls, but you should insulate the ceiling. If the basement is heated, it requires insulation in the walls, but not in the ceiling.

If you have a finished basement, open up the electrical outlets or pull off a small piece of the wallboard to see if there is any insulation behind them. You should have at least a 1″ thickness of rigid board in the walls or 3″ to 4″ thick batt or blanket insulation in the ceiling.

Insulation in crawl spaces and slab-on-grade construction is the most difficult to determine. If your house has an open crawl space, there should be either batt or blanket insulation between the joists. If the crawl space is closed, the insulation can be either between the joists or against the foundation walls, and it should be at least 6″ thick. If it is less than 6″, add more.

If your home is built on a slab, it is nearly impossible to look under it and see if there is any insulation. If you dig down beside the footings, you should find a rigid board around the perimeter of the slab that reaches below the frost line, which may be anywhere from 6″ to 2′ below ground level. But sometimes rigid board insulation is laid on top of the concrete and then sandwiched under the subflooring and floor. You can check this by drilling through the floor to the subflooring and into the insulation. As soon as you stop seeing wood shavings come out of the hole you know you are into the insulation. If there is no insulation, your drill will go through the wood subflooring and come to an abrupt halt against the concrete.

Heat Ducts and Hot Water Pipes

Hot water loses one degree of heat for every foot of its run through an uninsulated pipe. Air does not have quite so dramatic a heat loss, but it loses warmth quickly enough so that any extended run through bare ducts will cause an appreciable demand on your heating plant to burn more fuel than it should have to consume. Wherever the ducts or water pipes pass through an unheated portion of your house, such as the basement (where they probably begin) or the attic, they should be wrapped in the thickest duct insulation you can find; seal all of the joints between pieces of duct insulation to avoid any condensation.

Vapor Barriers

Many forms of insulation, particularly batts and blankets, come with a vapor barrier attached to them in the form of polyethylene sheeting, aluminum foil or an impregnated paper, which specifically resists water vapor from entering the insulation itself. Warm, moist air passing from a heated room through outside walls will condense as it strikes any cold surface. If it condenses on the insulation and dampens it, the insulation will lose practically all of its protective qualities. So the vapor barrier is used to protect insulation material, and when the insulation is installed, the barrier is always placed facing the warmest surface. Thus, it goes against the inside of any wall, or facing the room side of a ceiling. If you are making a new installation with batts or blankets, you need only make certain that the vapor barrier is facing the warmest surface. If you are blowing loose fill into an area, there are many instances when you can lay down sheets of 4-mil polyethylene first. Make certain the sheets are taped together at their joints and that any tears or holes are patched with duct tape before blowing loose fill over the plastic.

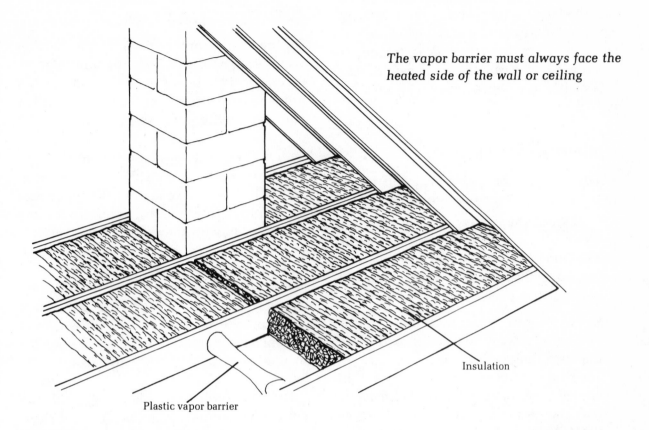

The vapor barrier must always face the heated side of the wall or ceiling

Insulation

Plastic vapor barrier

If there is already insulation in your home, but no vapor barriers, it is difficult to add a barrier. Should moisture become a problem, it will show itself as mildew, rotting, peeling paint and wallpaper (to say nothing of what all that moisture is doing to the insulation inside the walls). You can reduce the problem of peeling paint by putting an oil-based enamel under an alkyd-based top coat on the walls. Penetrating floor sealers or four or more coats of floor varnish will also act as a reasonable vapor barrier on floors. Floor wax applied to wood panels can also assist in retarding moisture-filled warm air from penetrating through to the insulation. Vinyl wall coverings and resilient flooring both make good vapor bar-

riers, although the flooring sheet goods are better than squares because they have fewer joints. The standard vapor barrier materials, the ones used in new construction, are 4-mil polyethylene sheeting, aluminum foil and various impregnated papers known as felt. All of these, when applied to a surface, will retard moisture.

DO-IT-YOURSELF—OR NOT

Most of the insulating that should be done in an average home you can do yourself. Most, but not all. If you elect to use loose fill that must be blown in place, it is less time-consuming and more reliable if you hire a qualified contractor who has both the experience and the proper equipment to do a competent job. Even including the cost of the contractor, insulation will pay for itself in a matter of months by saving on the cost of fuel.

HIRING A CONTRACTOR

When hiring a contractor to install insulation, follow this four-step procedure:

1 • Check out the contractors you are interested in with the Better Business Bureau. The bureau will give you references you can call and ask about recent jobs the contractors have done.

2 • Know what work you want done and the R-value you want installed. Ask the contractors to survey your home and give you a written estimate for the job that includes a full description of the materials they intend to use. (Get at least three estimates.)

3 • Be sure the contractor you choose has insurance covering his employees as well as any damage done to your home. Also try to get a written understanding of the way the finished job will look.

4 • During the work time, check the materials that are brought into your home to make certain you are getting the type of insulation that was agreed upon, or its equivalent. *Do not, under any circumstances, pay the contractor his final payment until the work is completed and you are thoroughly satisfied.* That doesn't mean you should wait a year to see if the insulation reduces your heating bills, but make certain that all of the agreed-upon work has been totally completed.

INSULATING ATTICS

During cold months, most of the heat in your home rises into the attic and then escapes outside. In hot summer months, an uninsulated attic collects so much heat that your air-conditioning system has to work overtime to keep the rest of the house cool. So if nothing else, insulate your attic.

Before you even consider the insulation you will need, make certain that your roof does not leak. Examine it carefully for any sign of daylight showing through or any water damage. In most instances, you can make your repairs by replacing a few shingles, but the areas around the vent pipes and chimneys may need recaulking, and if you discover too many problems, you will have to replace the entire roof. Do it before the insulation is installed because damp insulation is worthless.

Determine what you plan to do with the attic. If it is unfinished and will remain that way, it will require a different amount of insulation than if you intend to finish it. Whatever your plans for the attic, measure its length and width and multiply the two dimensions to get the total square footage.

Now measure the distance between joists; this will most likely be either 16″ or 24″ from the center of any joist to the center of the joist nearest to it. The blankets or batts that you buy should be the same width as the distance between the joists. How many batts or blankets you need is determined by multiplying the total square footage of the attic times .90 (if the joists are 16″ on center) or .94 (if the joists are 24″ on center).

You must have a moisture barrier in the attic, and the simplest way of getting one is to buy batts or blankets with a vapor barrier attached to one side.

Before you begin work, make sure there is adequate light. In most attics it is best to use a droplight you can hang from the rafters above wherever you are working and then take with you as you move around the space. With an unfinished attic, you should also have some pieces of ¾" plywood large enough to span the tops of the joists and serve as a movable floor. You can walk on the joists safely enough, but if you slip and step between them, you'll fall through the ceiling to the top floor of your house. If you are working with either rock wool or fiberglass, wear long-sleeved shirts, goggles, a breathing mask and above all, gloves. You will be walking around under low rafters and a roof that probably has a lot of nails sticking through it, so wearing a hard hat is a good idea as well.

Laying Batts or Blankets

Keep the vapor barrier down, facing the ceiling attached to the underside of the joists. Start under the eaves and, with a stick, push the end of the batt or blanket into the corner between the joists and the rafters. Then press the insulation firmly down between the joists without compressing it.

Continue working from the eaves toward the center of the room, then go to the opposite side of the attic and start from the eaves

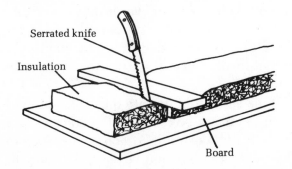

How to cut blanket or batt insulation

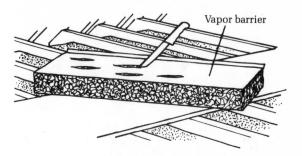

Make long slits in the vapor barrier if a barrier already exists in the insulation being added to

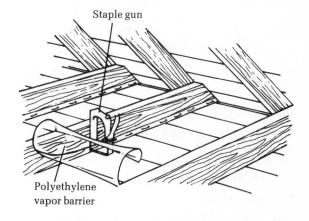

Staple 4-mil polyethylene to the joists to act as a vapor barrier before installing loose fill insulation

again. Whenever you must cut the insulation to go around obstructions, use a sharp serrated knife; it is easiest if you lay a board along your cutting line and compress the material as you cut it. Trim the insulation so that there is approximately 3" of clearance around any lights, vents, fan motors, or other heat-producing equipment.

If there is already insulation in the attic and you are merely increasing its R-value, the procedure for laying batts and blankets is the same. The only difference is that you should use insulation that does not have a vapor barrier attached to it. Chances are the existing material has a vapor barrier of its

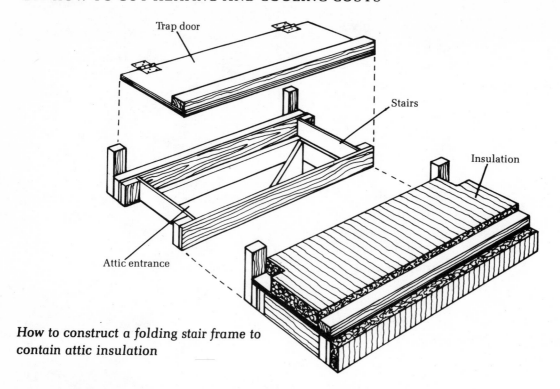

Trap door

Stairs

Insulation

Attic entrance

How to construct a folding stair frame to contain attic insulation

own, and if you add more insulation with another barrier, moisture will eventually become trapped between the two barriers.

If the only insulation you have is with a vapor barrier, slash holes in the barrier with a sharp knife before you lay it in place.

Applying Loose Fill

If you elect to use loose fill in an attic floor, you first have to staple sheets of 4-mil polyethylene to the sides of the joists to create a vapor barrier. Lay the strips of polyethylene between the joists and fold the edges up against the wood; then staple them in place every few inches.

Next, calculate the square footage of the area to be insulated (length × width × .90 or .94) and talk to your insulation dealer. He has a chart that will tell you how much coverage each bag of loose fill will give you to

achieve whatever R-value you wish. When you are buying the fill, also buy a batt or two that can be cut up to go around any vents or heat-producing appliances to keep them from being covered with the loose fill. Install the protection batts before you begin filling.

Pour the loose fill from its bag onto your vapor barriers, allowing it to mound between the joists. Use the back of a garden rake or a board to even the insulation; make certain you leave no voids anywhere between the joists. Work from the eaves toward the stairway, and finish by stapling a batt of insulation to the attic side of the stair door.

If the stairs fold up into the attic, you will have to erect a box around the stair hole to contain the loose fill. Attach a trap door to the top of the box and cover it with a batt.

If you are adding loose fill to existing insulation, do not put a vapor barrier down, merely pour the fill.

Insulating an attic that is finished, or about to be finished, requires insulation in both the rafters and the walls, but not the floor. If you are planning to heat the attic, remove all insulation between the floor and ceiling of the top floor. Insulation is designed to form a barrier between heated and unheated areas; between heated areas it serves to inhibit the flow of air through your house.

Blankets or batts should be stapled between the rafters before they are covered with paneling or wallboard. Begin by pushing the end of the material against the ridgepole and then staple it every 8″ to both

rafters. If there is a ceiling over the attic room, the insulation should be stapled between the rafters before it is covered with wallboard. Remember that the vapor barriers always face the heated room.

Beyond the room, between the eaves and the attic room walls, insulation goes into the attic floor, and, of course, the outside walls of the attic must also be filled.

During the course of installing all this insulation, you are bound to encounter tight corners and irregular places that are too small for the 16″- or 24″-wide batts and blankets. When dealing with irregular places, cut the batt or blanket 1″ wider than the space it is to fill, then peel ½″ of insulation off each side, leaving a small flange along each edge of the batt that can overlap the

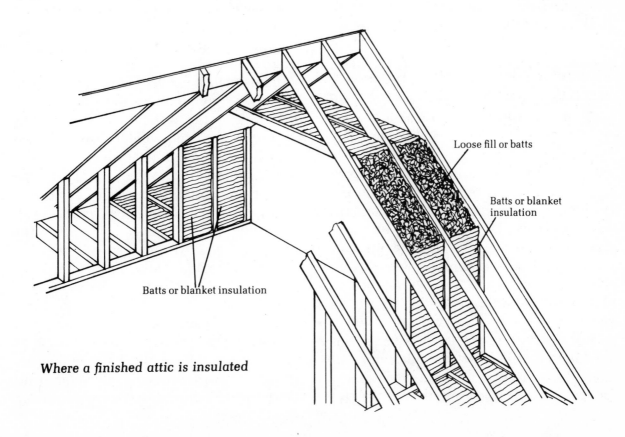

Loose fill or batts

Batts or blanket insulation

Batts or blanket insulation

Where a finished attic is insulated

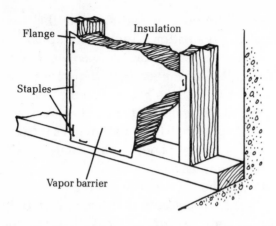

Cut ½" of insulation off each edge of the batt or blanket and use the vapor barrier as stapling tabs

joists or studs and be stapled to the wood.

Finished attics present the problem of completed walls and floors which may not be easily opened. In this case, your only viable alternative is to hire a contractor to blow loose fill into the uninsulated spaces.

INSULATING EXTERIOR WALLS

The second most important part of your house to be insulated is its exterior walls. All of them: front, back and sides. In many older houses, if you can get into the attic you may find access to the spaces between wall studs. Simply fill them by pouring loose fill into the holes. If you do not have this type of construction, your only recourse, short of completely opening up all four walls, is to drill holes through either the inside or outside of the walls every 16" and blow insulation into the cavities. Even so, you will not have a vapor barrier, which means you will have to paint a barrier on the interior walls. For example, you can apply two coats of a good alkyd-based semigloss paint over a primer coat of alu-

minum paint, although two coats of aluminum-in-varnish paint will form an effective vapor barrier all by itself.

You can always call in a contractor to fill the walls in your house or you can rent insulation blower units from tool rental companies and large insulation dealers. If you elect to take on the blowing yourself, the first order of business is to drill small holes into each wall every 16" between every pair of studs. If you are working from the inside of the walls and there is a crown molding along the ceiling, remove it and drill so that when the molding is replaced, it covers the holes. Once the holes are drilled, lower a weighted string down the inside of the wall to make sure the cavity is open all the way to the floor. It may not be: there are often spacers and fire-stops placed between studs. If your weight encounters an obstruction, a second hole must be drilled in the wall below it. You will also find 2"×4" fire-stops around windows and doorways.

Unless your home has a masonry exterior or metal siding you have the alternative of drilling into the walls from the outside. Both clapboard and shingles can be pried away from the exterior sheathing and holes drilled behind them. Begin by prying up the top course of either shingles or clapboard, then peel back the building paper that is directly under it. You may have to slit the paper horizontally, then vertically at each end before it can be folded down from the sheathing it is nailed to. The sheathing is typically ¾"-thick plywood or, in older homes, tongue-and-groove boards nailed to the studs, which in turn stand either 16" or 24" on center. With an outside wall they are usually 16" apart. Drill holes along the eaves line every 16" and lower a weighted string

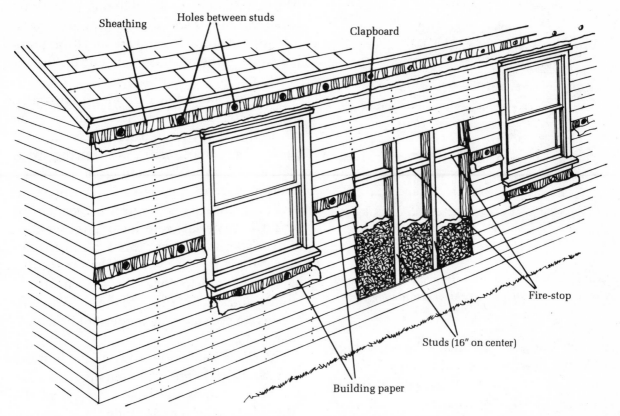

Where holes must be drilled for blowing in insulation

into each of them. If the house is more than one story high, the string will most certainly stop at the top floor, although it may encounter spacers or fire-stops before that. In that case, you will have to drill another hole below the spacer until you have enough holes to fill the entire cavity from the foundation to the roof.

When the cavities have been filled with insulation, each hole must be plugged. You can nail tin can lids (or bottoms) over them, or use specially designed plastic plugs which snap and lock into the hole, or cut wooden plugs for them. Then staple the building paper back in place and replace the shingles or clapboard. When filling holes in interior walls, you can either stuff them with paper or insulation batting, then fill them with plaster of paris, spackle, or wallboard compound and paint them. Most likely the walls will have to have their vapor barriers painted on anyway.

INSULATING BASEMENTS AND CRAWL SPACES

Basements

If you don't intend to use your basement for anything except storage, all you really have to do is insulate the ceiling. The easiest way to do this is with batts or blankets nailed

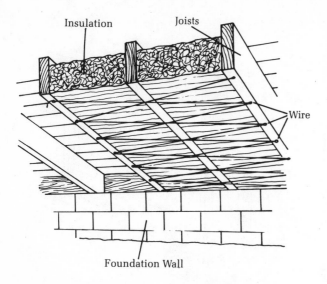

Insulation in a basement ceiling must be retained by wire or lath strips to keep it from sagging

between the joists. Estimate the amount of material you will need, and when you buy it also get some chicken wire, lath strips or heavy-gauge wire to hold the insulation in place and keep it from sagging. Face the vapor barrier up against the ceiling and staple the batts or blankets between each pair of joists. Then either nail strips of lath across the joists, staple wire mesh to them or lace wire back and forth between nails driven into their undersides. However you do it, the objective is to keep the insulation from sagging under its own weight.

If the basement is used as a workshop or living space of any kind (such as a family room), it should have insulation on its walls but not on the ceiling. If the walls are already finished, the easiest method of insulating them is to drill holes and blow loose fill in between the studs. If the room is still under construction and the studs are exposed, you can use batts, blankets or rigid

board; but before you install any insulation, make certain there are no problems with ground moisture. If there are moisture problems, solve them *before* you put up the insulation.

You can stand studs up along the walls, fit batts or blankets between them, then cover the studs with paneling or drywall panels. Or, you can nail 1″×2″ furring strips to the masonry walls and glue 1″- or 2″-inch-thick rigid boards to the strips, then cover the boards with drywall panels or whatever materials your local building code allows you to use over this kind of insulation. No matter how you insulate the walls, be very careful to cover the sill. In fact, if you do nothing but insulate the sill, you will preserve a considerable amount of heat each year.

The sill is bolted to the top of the foundation walls and forms the outside joists on two walls and is the header for all of the other joists at both ends along the other two walls. It is above the ground and is suscep-

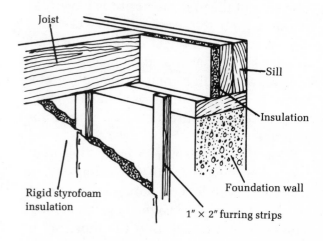

Rigid board insulation can be stapled to 1″ × 2″ furring strips

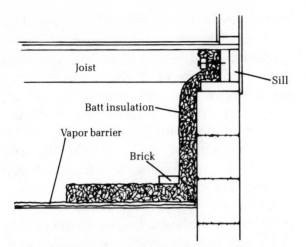

Insulation under a crawl space

tible to both moisture and heat loss. Stuff batts or blankets against the sill all the way around the perimeter of the house basement and staple it in place, with the vapor barrier facing you.

Crawl Spaces

However you do it, insulating a crawl space is mean, thankless work. In a heated crawl space that cannot be used for anything, simply staple insulating batts to the flooring underlayment and let them hang against the walls. Hold each batt in place with bricks tucked into the corner between the ground and the wall. The ground must have a 4-mil polyethylene vapor barrier spread over it and held down with bricks.

INSULATING SLAB FOUNDATIONS

In theory, if you have a slab under your house or any part of it (under the garage, for example), the outside of the slab was cov-

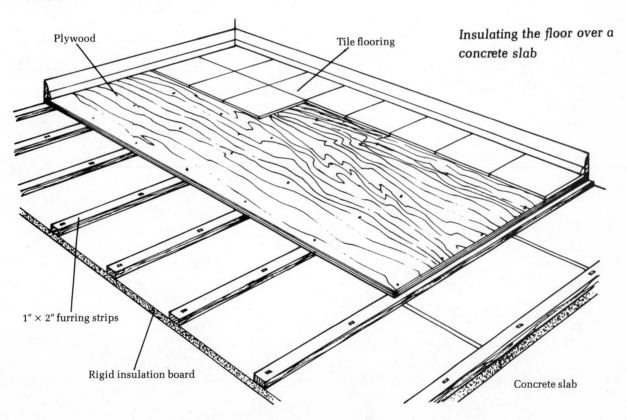

Insulating the floor over a concrete slab

ered with rigid insulation boards at least as far below ground level as the frost line, which could be as deep as 2'. The rigid boards were glued to the concrete slab with a mastic and then covered with stucco, plaster, building paper, or whatever your local code approves as a covering for polyethylene insulation. If the builder did not insulate the slab, you can do it yourself, but the job requires a considerable amount of digging, even though it will result in an appreciably warmer slab floor.

You might not want to bother insulating a slab under your garage—unless you plan to convert the area into living space, in which case you can also insulate the garage floor with rigid sheets. Apply a coat of vapor-repelling paint to the slab floor and then glue rigid sheets of insulation to the concrete, using whatever mastic is recommended by the insulation manufacturer. Next, nail 1"×2" furring strips, spaced 16" on center, to the rigid sheets. You will need to use masonry nails, which are driven through the rigid sheets and into the concrete. When the strips are in place, nail ½" exterior-grade plywood sheets to the furring strips and put whatever floor you intend to have on top of that. The floor will stay relatively warm and dry for years to come.

3 insulating WINDOWS and DOORS

WEATHER STRIPPING, CAULKING AND VENTILATION

The National Bureau of Standards estimates that 40% of the energy used in the average American home is totally wasted (along with the cost of fuel needed to create that energy). But full insulation can retrieve only three-quarters of that 40%. The second major culprits of wasted energy in practically every home are the doors and windows. Especially the windows. Government and industry studies have shown that the doors and windows of your home can be responsible for as much as 40% to 50% of your home's total heat loss, which amounts to a straight waste of approximately 20¢ out of every dollar you spend for heating or cooling fuel.

You will never get any door or window to be as energy efficient as a well-insulated wall, but by weather-stripping and caulking each of them, you can dramatically reduce the cost of heating or cooling your house. Neither weather-stripping nor caulking is difficult to do, given the modern products available at almost any hardware store or building supply center, although to do both correctly does require a modest amount of working time.

The ⅛"-wide crack around the perimeter of an average exterior house door is equal to a 6"-square hole in the middle of that door, and will allow an equal volume of warm air to escape from your home every minute of every hour of every day in the year. Since warm air is constantly traveling from warm surfaces toward colder ones, it is not only leaving your home all winter, but entering it all summer. As a result, both your heating and cooling systems must work longer and harder to keep your home comfortable.

Windows are even worse villains than doors because glass happens to be an excellent heat conductor. Its ability to allow heat to pass through it can be inhibited by installing either double or triple panes or, at the very least, storm windows. But no matter what kind of windows you have, if they are movable there are cracks around the sashes and at the joints between the window casing and the frame of the house. The tiny cracks that annually appear around window and door frames can be sealed with caulking; the larger spaces between window sashes or doors and their frames can effectively be blocked with weather stripping.

WEATHER STRIPPING

You can evaluate whether a door or window needs to be weather-stripped simply by standing in front of it and peering along its edges. If you can see daylight, you need to add weather stripping. If you cannot see daylight but can feel cold air when you hold your hand near the crack, you still need weather stripping. Go around your house and feel along the edges of every window and door. Some of the cracks may appear to be airtight; others will not. But when you add up all the offending spaces, you will probably arrive at the conclusion that you have a very drafty home. Fortunately, there are a variety of effective weather-stripping products that can be applied to doors or windows with relative ease.

FELT

Felt strips are made in several widths, thicknesses, qualities and colors, and are usually tacked in place, although some versions come with an adhesive backing. Felt strips used to be nailed along any offending crevices, but that is not where they are most effective; and besides, if you put them around

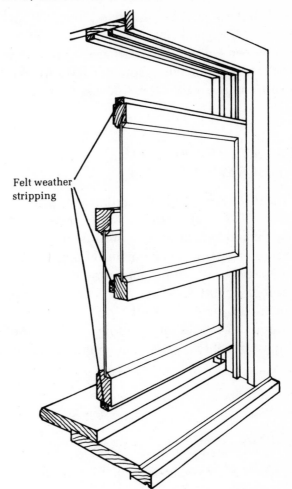

Felt weather stripping

Where to attach felt strips on a double-hung window

a door or window sash, they tend to be unsightly. But if you attach felt strips to the bottom of a lower window sash or to the top of the top sash, as well as to the inside bottom edge of the upper sash, you will have an effective kind of gasket that is not often visible. You can also tack or adhere the felt to the stops in a doorjamb.

How to Attach Felt

1 • Cut a felt strip, with scissors, to fit the surface.

2• Tack one end of the strip, then nail it every 2" or 3", keeping the felt taut as you proceed along the surface. Since felt stretches somewhat, you may have to trim off some excess after the last tack is in place.

PRESSURE-SENSITIVE ADHESIVE-BACKED FOAM

This is the easiest of all weather stripping to apply. The foam can be either rubber or plastic and is sold in rolls of different thicknesses and widths. The material is good for one or two years of use and should never be painted (paint hardens the foam and reduces its resiliency). It can be applied around the edges of windows and to the stops on doors: When it is compressed, it will effectively seal out air. But you can only use it in areas that do not have to withstand friction, such as the bottoms or tops of window sashes or against the doorstops, since any sideways pressure against the strip will break its bond. When weather-stripping a door with foam, attach the strips on the hinge side to the jamb; on the lock side, attach to the doorstop.

How to Attach Pressure-sensitive Adhesive-backed Foam

1• Try to work on a warm day, with the temperature at least 60°F. Clean the surface to be weather-stripped with a good detergent to remove all grease and dirt. If you are replacing adhesive-backed weather stripping, make sure all the old adhesive is removed, and the surface is dry.

2• Cut the strip with scissors to fit the surface and peel back an inch or two of the backing.

3• Press one end of the strip in place, then continue along the surface, peeling off the backing a few inches at a time and pressing the strip into position. If the strip refuses to adhere to the surface, paint a coating of contact cement on the back of the strip and on the surface. Allow the cement to dry for 20 minutes, then apply the weather stripping. Be careful to align the strip properly before the two cement-coated surfaces meet; contact cement bonds instantly and is almost impossible to separate without tearing the weather stripping.

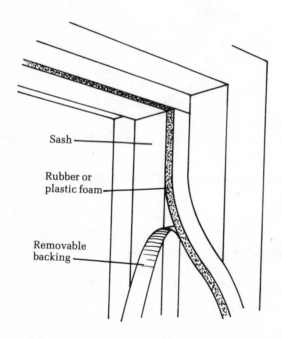

Sash

Rubber or plastic foam

Removable backing

When applying pressure-sensitive adhesive-backed foam, peel off only a few inches of the backing at a time

SERRATED METAL

This type of weather stripping can be either

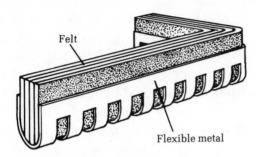

Metal-backed felt is sold in coils

felt or vinyl-backed with a strip of metal and comes in long rolls. Because it provides the strength of metal plus the ease of application of felt, it has all but replaced felt for many uses. Serrated metal is attached in exactly the same way as felt, but you will need a pair of tin snips to cut the metal backing.

TUBULAR GASKETS

These are made of very flexible vinyl and may be hollow or have a foam core. They are unsightly, so they are usually installed on exterior surfaces, but they offer the great advantage of fitting into oddly shaped spaces. You can attach a tubular gasket almost anywhere; position it so that its tube portion is against the crack you are sealing. Tubular gaskets can be used around the edges of doors as well as windows but should never be painted, since paint will stiffen them. The foam-filled versions hold their shape better than the hollow tubes.

How to Install Tubular Gaskets

1 • Measure and cut the gaskets to size.

2 • Position the strip and nail one end of it in place.

Tubular gaskets may be hollow or have a core

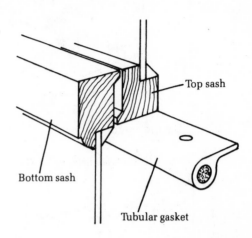

3 • Pull the gasket taut and nail it every 2″ or 3″.

SPRING METAL STRIPS

Spring metal strips are primarily used around the sashes in wooden double-hung windows and for weather-stripping doors. They can be either flat or V-shaped, with copper, bronze, aluminum or stainless steel finishes, and are sold in rolls which are nailed or held to a surface by an adhesive backing.

How to Install Spring Metal Strips to Windows

Metal strips are cut with tin snips to fit each sash track and should be about 2″ longer than the sashes so that they are exposed when the window is closed. As with all vertical installations in the sash channels, the flared flange should face the *outside* of the window. Strips mounted to the top of the

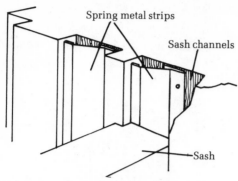

Spring metal strips

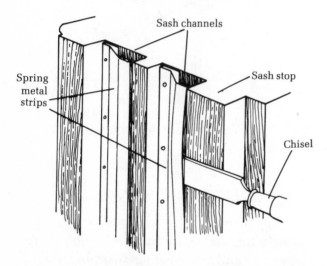

After the spring metal strip is installed, carefully pry it open to improve the air seal

upper sash and to the bottom of the lower sash are also mounted with their flared flanges to the outside. The strip on the bottom of the top sash is aimed *downward*.

1 • Cut all the strips to size and position each one before you install it, to make sure it is not interfering with any hinge, lock, sash pulley or other hardware on the window. Trim the strip wherever it may cause interference and cut it at a 45° angle (miter) at any corner where the horizontal and vertical strips will meet.

2 • Position each strip carefully and nail it at both ends of its nailing flange without driving the nails home. Then tack the center and all along the strip without driving the nails home. The nails must be driven flush with a nail set so that the strip is not damaged by hammer dents.

3 • With a screwdriver or chisel blade, carefully pry open the unnailed edge of the strip to make it fit snugly.

If you are installing metal strips with adhesive backing, cut and position each strip as described in steps 1 and 2. Then stick the metal in place instead of tacking.

How to Install Spring Metal Strips to Doors

Spring metal is a popular weather stripping for doors because it is not visible when the door is closed. Manufacturers sell spring-metal weather stripping kits for doors which include a small triangular piece that is installed next to the striker plate so that air cannot pass through the lock area of the door. Strips are attached to the top and side edges of the doorjamb.

1 • Cut the side pieces and nail or stick them in place.

2 • Miter and fit the top piece across the top of the jamb.

3 • Flare out the edges of the strips with a screwdriver.

4 • The lock side of the door, which has the striker plate, has two jamb strips that end just above and below the striker plate. The triangular piece is attached at right

angles to the plate and fits snugly between the ends of the jamb strips.

CASEMENT AND JALOUSIE WINDOW GASKETS

These are U-shaped tracks made of vinyl, which are cut with scissors or a knife and fitted over the edges of the glass or casement. The ends of each strip should be cut at 45° angles to form mitered corners. It is not necessary to glue or otherwise hold them in place.

THRESHOLDS AND DOORSWEEPS

Doorsweeps can be made of wood and felt, wood and foam, aluminum and vinyl, or spring steel. The spring metal is mounted on the outside of a door opening inward; all other versions are attached to the inside of the door and have a strip of foam or felt at-

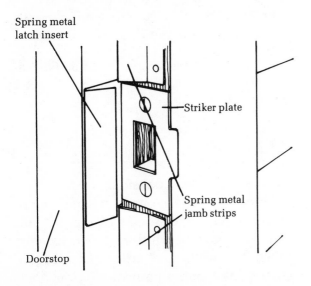

The triangular piece of spring metal is positioned against the doorstop

tached to them which hangs below the door and fills the gap between it and the threshold.

Doorsweeps are designed to fill the space between the bottom of a door and its threshold. The threshold is installed between the

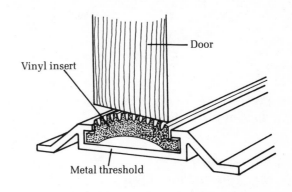

Side view of an aluminum threshold

doorjambs and is usually a piece of shaped oak, which has no particular insulating value. In fact, the wooden threshold is normally a fraction of an inch below the bottom of the door, leaving a gap for warm air to escape outside. Since the threshold is nothing more than a board, it is easily pried off the floor with a claw hammer, and can then be replaced by an aluminum substitute. Most aluminum thresholds have a vinyl insert which pushes against the bottom of the door to create an airtight seal when the door is closed. There are also interlocking versions, but to install these, the door must be removed and its bottom edge rabbeted to accept one of the pieces. While the interlocking systems are efficient, they are also difficult to install, as they must be fitted perfectly; it is easier, and just as energy-efficient, to install any of the one-piece aluminum units.

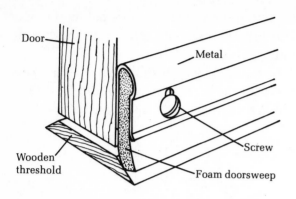

Doorsweeps are usually attached to the inside of the door, at the bottom

How to Install a Replacement Threshold

1•Remove the old threshold with a pry bar or hammer. If the threshold is pinned at each end under the doorstops (or sometimes under the jambs themselves), saw through it with a hacksaw (the wood is only about ⅝" thick). The saw should be started with its blade flush against the doorstop and kept parallel with the floor so that it does not have a chance to scratch the flooring around the door. If you are replacing a metal threshold, you may have to remove the vinyl insert and undo the retaining screws hidden in the groove.

2•Clean the wood under where the new threshold will sit and lay it in place. If it is too long for the width of the door, trim it with a hacksaw. Drill pilot holes for the retaining screws and screw it in place.

How to Install Sweeps

If, after installing the threshold, there is still a gap between the threshold and the door,

or if you elect not to replace your old threshold, you can attach any of several kinds of doorsweeps to the bottom of the door. Most doorsweeps are tacked or screwed to the inside of the door.

1•Cut the sweep to the exact width of the door.

2•Tack or screw both ends of the sweep in place, then continue fastening it every 2" or 3". Some doorsweeps attach to the outside of the door and are designed to flip up and down as the door passes over its threshold. These are installed in the same way as any doorsweep, except they are attached to the *outside* of the door.

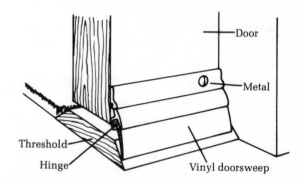

Some doorsweeps are designed to go on the outside of the door and flip up and down as the door goes over the threshold

CAULKING

The average home develops enough tiny cracks between the joints of dissimilar materials to equal a foot-square hole in the side of your house. So even though you have insulated the walls and attic, and weatherstripped the moving parts of all your doors and windows, you still have not completely sealed your house.

Metals expand and contract with heat and cold at different rates and to different degrees. Wood swells and shrinks with every change in the humidity. On a cold day, the wood (including the furniture) in a room heated by a steam radiator will change its size as often as every fifteen minutes! When any of these various materials are butted together, the joint between them can open and close with amazing regularity and to amazing dimensions, depending on the temperature, moisture, or pressure exerted on them.

Not only do all of these cracks offer avenues of escape for warm air, but moisture can seep into them, causing wood to rot and insulation to lose its effectiveness. They become doorways for insects entering your house. They can, in time, begin to literally destroy the home you live in. Thus, mankind invented caulking, which, in most cases, can resolve, or at least retard, the problem of cracks. Caulking has been around ever since man set out to sea in boats. For centuries, tar and oakum were our only caulk. Oakum is actually a piece of hemp treated with tar to resist moisture and meant to be stuffed into any large crack and then covered with a caulking compound. You can still buy oakum, and if you are confronted with any crack that is more than ½" wide, you can fill it with oakum.

WHERE TO CAULK

Anywhere on the outside surfaces of your home two separate pieces meet, there is a crack. If the materials of the two pieces are different, sooner or later that open space should be caulked. Most caulks should be applied when they are warm, so arm your-

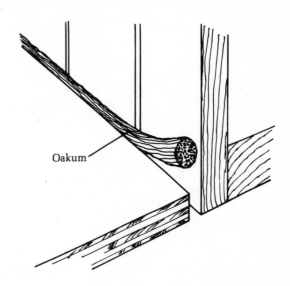

Oakum is the oldest form of caulking known

self with a caulking gun on a warm day and inspect the outside of your house, looking for places to caulk. Pay particular attention to these areas:

1 • Anywhere the frame of a door or window joins the side of the house.

2 • The sill where the top of the foundation meets the side of the house.

3 • Anywhere steps and porches meet the side of the house.

4 • Around the base of chimneys, vent pipes or anything else that sticks out of your roof, including between the edges of any flashing and the roof shingles.

5 • Around the perimeter of any plumbing or exhaust pipes that come through the walls of the house. The pipes usually have a flange around them. Pry the flange away from the wall and caulk behind it, then push the flange back in place and caulk around its perimeter.

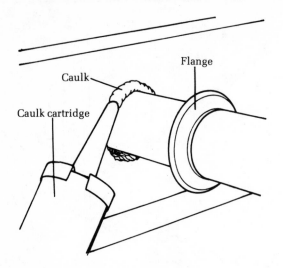

Caulk behind the pipe flange, as well as around its outside

6•The corner seams where the siding meets itself.

7•The spaces between air-conditioning units and window frames.

HOW TO USE A CAULKING GUN

There are different caulks sold for different purposes, and each is applied a bit differently. So whatever product you use should be applied exactly according to the instructions printed on the package. In general, whenever you are using a caulking gun, bear these suggestions in mind:

1•Clean the area you intend to caulk. Scrape off any peeled paint and dig out all of the old caulking. Then wash the area clean of all dirt, wax or oils.

2•Try to work on a warm, but not too hot, day. If the outside temperature is below 40°F, warm the tube of caulking before using. If the day is over 80°F, the caulking will be runny; put it in the fridge for a few minutes before using.

3•Open a cartridge of caulk by slicing off the end of its spout with a utility knife or razor blade. Cut the tube at an angle and far enough up the tube so that the bead of caulk will be large enough to overlap both sides of the crack.

4•Insert the cartridge in your caulking gun and hold the gun at a 45° angle as you apply the caulk.

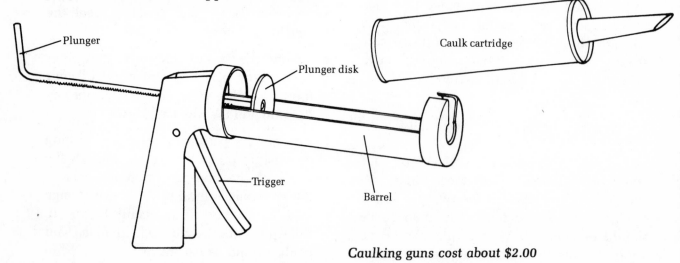

Caulking guns cost about $2.00

5•Many caulks can be pushed into a crack after they are applied over the area. But read the instructions on the cartridge; some require a metal blade, others a damp finger, and some cannot be "tooled" at all.

6•When you stop caulking, material will continue oozing out of the cartridge spout unless you disengage the plunger of the caulking gun. You can seal the cartridge by inserting a machine screw into the hole at the end of the spout.

TYPES OF CAULK

Caulk is sold under many brand names these days, and new ones come onto the market every year. So far, they all fall into one of eleven categories:

Oil-based•These are the least expensive caulks and will stick to most surfaces. But do not use them on moving joints as they eventually harden. They're best used on wood that has been given a coat of primer paint. Oil-based caulks shrink as they harden, but their usefulness can be extended by painting them as soon as they are hard enough not to be tacky. They can last for as long as seven years, but anything beyond one year is likely to be borrowed time.

Latex-based•These shrink less than the oil-based caulks and are guaranteed to last as much as ten years. In fact, some manufacturers promise twice that long. Latex hardens within half an hour after application, and if used outside, should be painted. Although latex caulks will stick to almost anything, they do a better job on metal or porous sur-

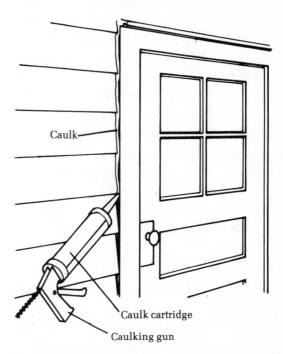

Caulk

Caulk cartridge

Caulking gun

When caulking, hold the gun at a 45° angle to the surface of the crack

faces if they are applied over a coat of primer. Apply latex in an extra-large bead to allow for its shrinkage; you can remove the excess with a damp cloth before it sets. After it has hardened, latex must be cut or peeled off whatever surface it has adhered to.

Butyl rubber•Butyl rubber caulks are usually the choice when sealing metal or masonry. They are not good in moving joints. They will last for as long as ten years and can accept any kind of paint. Butyl rubber caulks dry between 30 and 90 minutes; use paint thinner as their solvent/cleaner.

Silicone seal•Silicone will adhere to almost everything and shrinks very little. In fact, after it has cured, it will stretch up to seven times its width, making it ideal for moving joints, and will last for as long as 20 years. It dries tack-free within an hour, but does not cure for another two to five days. Clean silicone with naphtha or paint thinner before it is dry, otherwise you will have to cut it away. Some silicones cannot be painted.

Nitrile Rubber•Caulks of nitrile rubber will last between 15 and 20 years, but they shrink considerably and should not be used on any moving joints or in any cracks that are wider than ¼″. Nitrile does not adhere well to painted surfaces, but is extremely good for any high-moisture area. It dries within 20 minutes and will accept most paints.

Neoprene Rubber•These will last between 15 and 20 years and are particularly good on concrete. Neoprene does not shrink much and can be applied to moving joints if they do not move more than a ¼″. It dries tack-free in an hour but requires as much as two months to cure. Neoprene will accept most paints, but can only be cleaned or removed with such solvents as tolulene.

Polysulfide•These caulks will not shrink, last for more than 20 years, and can be used on moving joints. However, you must apply the polysulfides over the primer recommended by their manufacturer, and they are toxic for the three days it takes them to cure. The only cleaning agent is tolulene.

Hypalon•Not widely distributed, hypalon caulks are excellent on moving joints and will last up to 20 years. Hypalon will adhere to any surface and is easy to work with.

Polyurethane•Caulks of polyurethane can be used on moving joints and will last for 20 years, no matter what the weather may be.

They dry in 24 hours but need two weeks to cure. Polyurethane can be cleaned with acetone or paint thinner.

Polyvinyl acetate (PVA)•You will find PVA caulk in many stores, but no matter what its advertising says, only use it for small holes indoors. It loses its flexibility as soon as it dries.

Rope caulk•These will last for only a year, or perhaps two, and should not be considered a permanent solution to air leaks. They are pressed into a crack but do not bond to any surface and cannot be painted. They are often used around storm windows and are removed when the windows are taken off in the spring.

VENTILATION

While insulation, weather stripping and caulking are primarily installed to keep warm air from escaping from your house during cold times of the year, they also prevent hot air from rushing into your home during the hot months, causing your cooling system to work overtime. Nevertheless, a well-insulated, airtight home cannot stop hot air from entering and rising through it. It only slows it up. The air inside your attic on a hot summer day is cooked mercilessly as the sun beats down on the shingles. Heat radiates down from the underside of the roof and turns the insulation in your attic floor into a stovetop as it works its way down to the living spaces below. Without any form of ventilation in the attic, the temperature under your roof can top 150°F anytime it is over 95°F outside. And the rooms below the attic will go over 100°F. If you turn on your air conditioner and keep it running all day, you can get the temperature inside your house down to 75°F. But how much will 12 hours of electricity cost?

During the winter, an unventilated attic is likely to develop condensation on the walls, the rafters and the insulation, at which point the damp insulation loses its effectiveness and will cause your heating system to work longer and harder to keep your home comfortable. If you are not sure whether condensation is occurring under your roof, go up to the attic anytime this winter and look at the rafters for signs of frost, dampness or droplets. If you find them, at least you know the rest of your house is so airtight that the moisture from your heating, cooking, bathing and refrigerating has no place to go so it is collecting under the roof. On the other

hand, if you allow that moisture to collect for too long, your insulation will not only lose its effectiveness, but the wooden structural members at the top of your house will rot and have to be replaced.

The solution to both too much heat in the summer or moisture in the winter is to ventilate your attic, with either a static or power-operated ventilating system. Ventilators are not difficult to install and can be done by any homeowner. What may be a little more complicated is deciding what kind of vents you need and where to place them.

HOW MANY VENTILATORS?

Ideally, there should be ten air changes in your attic every hour so that warm air can

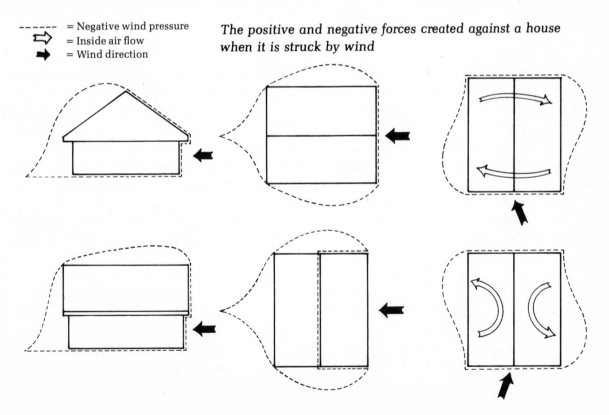

– – – – – = Negative wind pressure
⇨ = Inside air flow
➡ = Wind direction

The positive and negative forces created against a house when it is struck by wind

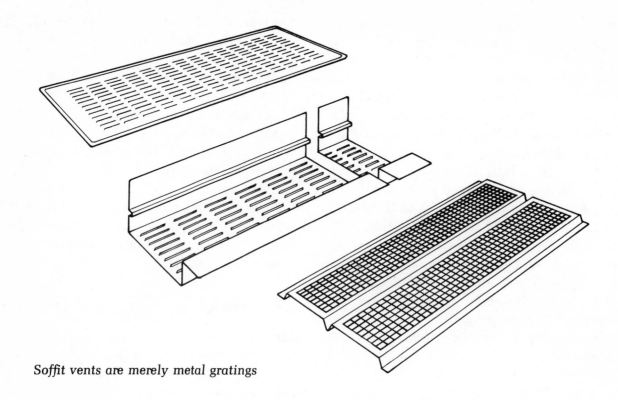

Soffit vents are merely metal gratings

escape in the summer and moisture does not have a chance to collect in the winter. To create this air flow, engineering tests have established that you need one square foot of ventilation (that is, a hole) for every 150 to 250 square feet of attic space. By multiplying the length and width of the attic and dividing by 150 and then 250, you can arrive at the minimum and maximum numbers of ventilators your home needs.

PLACING VENTILATORS

When the wind hits your house, it bounces back, creating a vacuum, or *negative* force, against the surfaces it has struck. The vacuum immediately forces air back against a different part of the house, causing a *posi-* tive force. You have to have vents in both places, so that the ones in a negative force area can allow air to be pulled out of the attic, while the ones in the positive force areas permit fresh air to enter under the roof. You must also remember that hot air rises and cold air sinks. So vents placed near the peak of the roof will allow warm air to escape from the attic, while vents situated near the attic floor will permit cooler air to enter.

Since the wind does not always strike your home from exactly the same direction, and since you also want to prevent rain, snow and even direct sunlight from entering the attic, a good ventilation system incorporates a whole series of carefully placed vents of varying designs.

The air flow in an attic having soffit vents

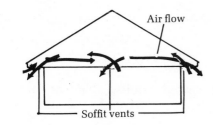

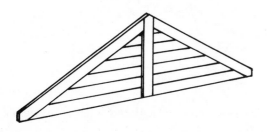

Gable-end louvers can be triangular, as shown here, or rectangular

TYPES OF VENTILATORS

Static Ventilators

Static ventilators are simply holes in the attic that allow air to flow in and out of your house. They are not particularly expensive, whether they are made of metal or wood. Nor is their installation any more difficult than cutting a properly sized hole in your roof or its eaves and nailing or screwing the vent in place. There are five basic static ventilator designs, each of which has a specific purpose and placement. Because their purposes and placements are specific, any good static ventilation system will incorporate at least two of the following five types:

Soffit vents • One of the vents you always use

to create a static ventilation system in your attic is the soffit vent. It is usually made of metal, and may be a small square grating or long, narrow rectangle. It is always placed horizontally under the eaves, near the attic floor, where the weather cannot enter it but cooler outside air can come onto the floor of the attic. Because the incoming air is cooler than the attic, it will not rise toward the roof, nor will it reduce the floor temperature very much, particularly if soffits are the only vents you have.

Gable-end louvers • These can be wooden or metal and may be either rectangular or triangular to fit under the peak of the roof in the gabled ends of your home. The louvers are angled downward so all but the most severe rain and snow is shed away from them before it can enter the house. Gable-end louvers are normally mounted in pairs, so they face each other from opposite ends of the house; they act as both intake and exhaust vents, depending on what the wind is doing. When the wind comes directly at one of the vents, it enters the attic and is sucked right back out again, with very little air change under the roof. If the wind is paralleling the gable vents, it is drawn in one vent, dips to the floor because it is cooler than the air in the attic, then is pulled out the other vent with a rate of flow that equals 70% of the outside wind velocity. How large the flow is, however, is limited by the size of the two vents.

Roof louvers • These are made of aluminum, wood, plastic or steel and are mounted near the peak (ridge) of the roof. They can be of any design that provides slits or screened openings that keep out the weather and in-

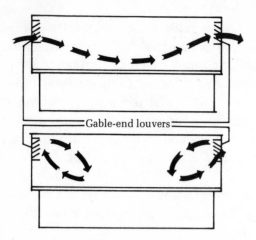

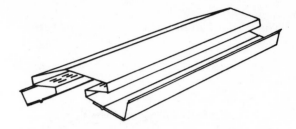

Ridge vents are installed along the full length of the house

If the wind comes directly at an end gable, the air enters the vent, turns right around, and goes back out again (bottom). When the wind parallels the end gable, it enters the attic, is drawn to the floor under the hotter attic air, and then goes out the other side of the house (top).

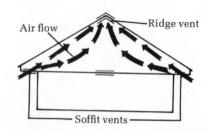

Air flow in an attic ventilated with soffit vents and a ridge vent

sects. Their major function is to exhaust the hot air hovering directly under the peak of the roof.

Turbines • These are a metal variation of the roof louvers and are topped by a multi-finned fan that rotates whenever there is enough wind to push it. The fan helps draw warm air out of the attic, but only the air immediately adjacent to it. Consequently, you need several of them lined up along your roof.

Ridge vents • These are made of aluminum and are placed over the ridge of the roof from one end of the house to the other. You can buy them in 4′ or 10′ sections and they will give you a net ventilation area of 18 square inches per linear foot. They are basically an exhaust vent that releases hot air from the full length of the house.

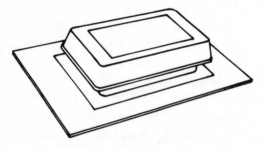

This is one of a variety of designs that roof louvers come in

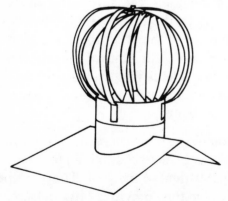

Turbines are roof louvers with fans that are turned by the wind

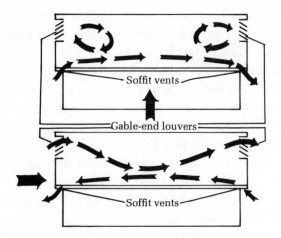

Air flow in an attic having soffit vents and gable-end louvers

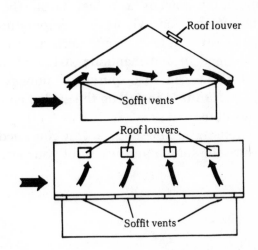

Air flow in an attic having soffit vents and roof louvers

Static Ventilator Combinations

Ridge and soffit vents • This is the number-one recommended combination of vents, and it offers the best possible air flow inside your attic. Cool air enters the soffit vents under the eaves, while the hot air rises toward the ridge vent at the roof's peak, where it is exhausted.

Gable-end louvers and soffit vents • Used together, these will create exactly the same air-flow pattern as if either type of vent is used by itself. Most of the air movement will be along the floor, and the temperature under the peak of the roof will remain as high as ever.

Roof louvers and soffit vents • The roof louvers will draw hot air outside, and if you have enough of them, most of the hot air can be removed. The soffits will allow cooler, outside air to wash over the floor. The problem is to achieve a balanced air flow and to do this you may end up with so many roof

louvers it would have been cheaper and easier to install a ridge vent. Nevertheless, the combination is better than just roof louvers alone.

Power Ventilators

Powered ventilating requires an electrically operated fan. You can install a fan in front of any gable-end louver, for example, and improve its exhaust capabilities. You can also position a fan in the floor of the attic over a vent that opens in the ceiling of your top floor. When turned on, the fan will draw cooler air from the living space, which in turn forces the hot air in the attic out whatever ventilators exist in the roof. There are also special power ventilators which are most often put in attics that have no other ventilation because they require few holes being cut into the eaves or roof and can be installed quickly.

A power ventilator is essentially a motor-driven fan attached to a ventilator that is installed on the slope of the roof near its ridge. Some units have a thermostat which automatically turns them on when the attic temperature reaches a preset level, then shuts off the fan when the temperature drops below that level. Some units are simply plugged in and manually switched on or off. You can also purchase powered gable-end units to install on the gabled ends of your house.

To determine how large a unit you need, measure the length and width of your attic and take the dimensions to the hardware outlet that sells powered ventilators in your locale. The outlet has ventilation charts that will help you compute the proper size unit to change the air in your attic ten times per hour. Buy the lowest wattage ventilator you can get that also delivers the highest efficiency of cubic feet per minute. The manufacturers of powered ventilators almost always provide detailed instructions with their products which explain exactly how to install the particular unit.

4 insulating WINDOWS and DOORS

THERMAL PANES, STORM WINDOWS/DOORS AND REFLECTIVE FILM

Even with your windows fully weather-stripped and caulked, heat will be conducted through the panes, because it is the nature of glass to do so; therefore, you will always lose heat through your windows. The game is to reduce that loss as much as you can. A single pane of glass ⅛" thick has an R-value of about one. When the temperature outside is 0°F, and your house is 70°F inside, the inside temperature of a single glazed pane window will be 18°F, or 14° below freezing! And all that warm 70°F air heads right for the coldest surfaces it can find, the window glass, which conducts it right outside.

Fortunately, one of the best insulators we have is air itself. And by placing two panes of glass ¼" or ½" apart and hermetically sealing the space between them, a window increases its R-value to around 1.5. If three panes of glass, separated by two air pockets, are used, the R-value jumps to 2.9. Thus, if the outside temperature is 0°F and the inside is 70°F, a double-glazed (paned) window has an inside surface temperature of 36°F. At least that is above freezing. A triple pane window has an inside surface temperature of 51°F, and that is quite acceptable. But don't start tearing out your windows yet; there are some economics to be considered.

If you have single-glazed windows and add a second pane of glass, you will raise the near floor temperature by 3° and reduce your heat loss by 50%. If you go from single-glazed windows to triple-glazed, the reduction is 65%. That extra 15% may not be worth the added cost of the triple-glazed

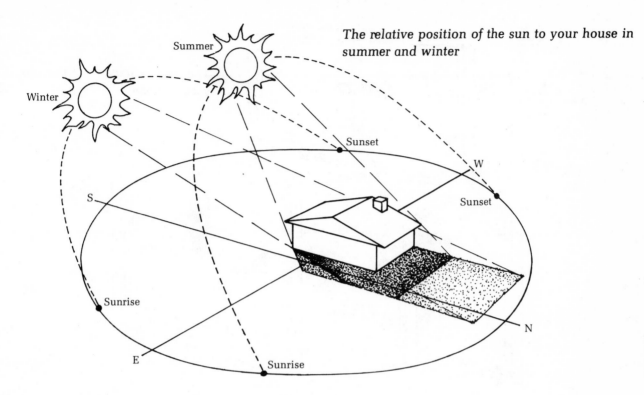

The relative position of the sun to your house in summer and winter

windows. Furthermore, you can come close to achieving the same 50% heat reduction over single pane glass simply by adding less expensive storm windows.

Double- and triple-glazed windows are known as thermal pane windows and the spaces between the panes may be hermetically sealed so that there will be no condensation between the panes. The ultimate in modern window efficiency is to have triple-glazed windows containing a vacuum between the panes. However, you will pay dearly for such an energy-efficient window, and it could be years before the reduced heating costs balance the high cost of replacing every window in your house.

If you are building a new home or an addition to your existing structure, the double- or triple-glazed windows should be your first choice. They can be purchased in any size for casement, sliding, double-hung, tilting or whatever permanent style you wish. They come in standard dimensions, or you can have them made to your specifications for not much more than the cost of stock sizes.

REPLACING A WINDOW

Removing the single pane windows you now have and replacing them with a multi-glazed unit is not a particularly complicated task, although each window will take you about half a day to install. However, replacing windows should not be your first do-it-yourself project. It takes some prior knowledge of carpentry, not only because of the

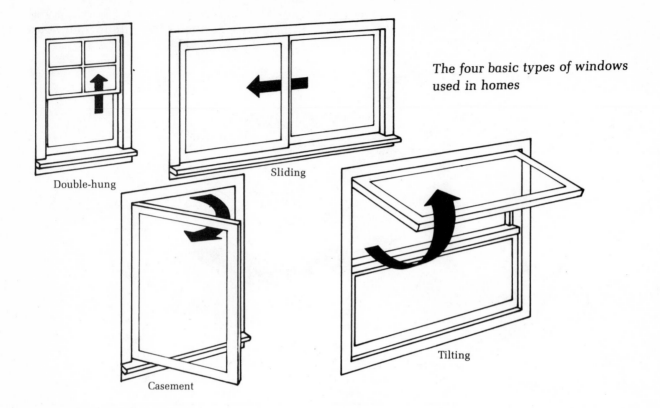

The four basic types of windows used in homes

Double-hung

Sliding

Casement

Tilting

basic steps involved, but because you may encounter some hidden problems, such as deteriorated wood around the window openings, which should be attended to.

The most complicated—and the most usual—replacement is a double-hung window, so if you can follow the steps given here, you can replace any type of window you might encounter.

1 • Measure the height and width of the window opening very carefully. In fact, measure it at least three times. You may discover that a stock size will fit the opening, but if the stock is not exactly the dimensions you need, have the new window made to order. There is very little difference between made-to-order and stock prices, so the only inconvenience is that you will have to wait two or three weeks for delivery. If you do elect to have your windows made to specification, in many instances a representative of the window company will want to measure them for you. Let him. Then if the windows don't fit you won't have to pay for them.

2 • When the replacement windows are delivered they will arrive fitted into their frames. They are completely assembled and all you have to do is position the frames in their openings. Begin removing the old window by prying off all inside moldings or stops. Work carefully so that the molding can be reused around the new windows.

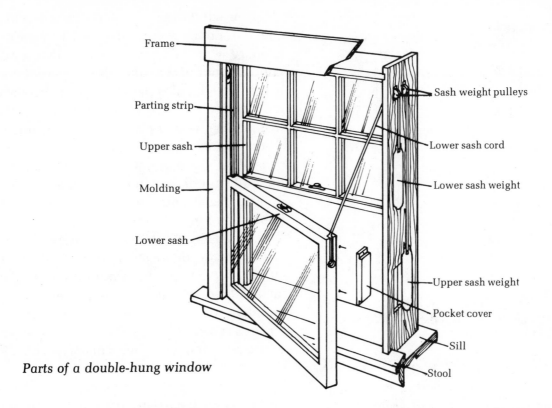

Frame

Parting strip

Upper sash

Molding

Lower sash

Sash weight pulleys

Lower sash cord

Lower sash weight

Upper sash weight

Pocket cover

Sill

Stool

Parts of a double-hung window

3• With the moldings removed, pull the lower sash out of its track and unhook the ropes or chains holding the sash weights. If you cannot get at the sash weights themselves, let them drop inside the frame. When the sash cords or chains are removed, lift the lower sash out of the frame.

4• Pry one or both of the center parting strips out of its groove with a chisel and pull the upper sash out of its track. Take off the sash cords or chains. Now clean the window frame of all obstructions. Pry any wood strips out of the channels with a chisel or screwdriver, undo any screws, pull out all nails, everything that might keep the frame from becoming four flat surfaces.

5• Unpack the replacement window. If it has a removable sash or sashes, take it out of the frame to make it both lighter and easier to install.

6• Put the new window frame in the wall cavity. It must be level and square in the opening. To make it that way, use shims wherever necessary. This is the time-consuming stage of any window installation because you have to have an absolutely square frame that is also level and does not bend or bow. When you are done shimming, pin the frame in place with a pair of screws in each side, and install the sashes. Make certain they work perfectly.

7• Take the window out of its cavity and run a bead of caulking around the inside of

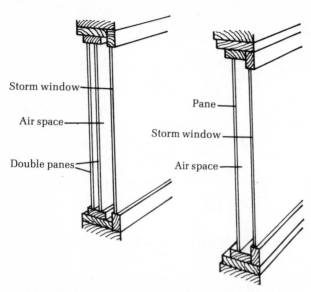

Double and single pane windows can improve their energy efficiency with the addition of storm windows

the blind stop. Caulk both sides and the top liberally, then position the window against the caulking, and secure it with screws. Keep trying out the sashes to make sure they work properly; you have to be very careful that the frame does not develop any bends in it that will inhibit their movement.

8 • Caulk around the outside of the frame. Also caulk around the inside before you replace the interior molding.

STORM WINDOWS

Tight, weather-stripped and caulked windows hold warm air inside during the winter, and keep it outside during the summer. Further protection against cold weather draining the heated air from your home is storm windows, which are replaced by screens during the warm months.

Storm windows, if properly installed and caulked, not only eliminate moisture on the primary windows, reduce outside sounds and dirt, but also reduce energy loss by as much as 50%. While the spaces between multiglazed window panes is ¼″ or ½″, the distance between a primary window and its storm window is more like 1¼″ or 1½″. Because that space is not vacuum sealed, it is not quite as efficient in reducing heat loss as a multiglazed unit, but it does a very creditable job of insulation.

You can purchase storm windows either custom-made or in stock sizes made with either wooden or aluminum frames. The metal frames are easier to handle and maintain than wood, but metal conducts heat better than wood. You can also purchase storm windows made with sheets of acrylic plastic rather than glass.

No matter what kind of single pane windows you have—double-hung, sliding, casement or tilting—if they are in reasonably good condition, they should have storm windows. With a sliding window, the storm window or panel can be attached to the outside of the window frame or fitted inside it. If you elect to put your storms on the outside they must also be weather-stripped; the storm window fills the frame, so obviously you cannot open the window for ventilation while it is in place.

Casement windows must have their storms attached to the inside of the sash, so these too cannot be opened until the storm window is removed. Tilt windows may have their storms clipped to either the outside or the inside sash, depending on which way the window swings open. If it opens outward, the storm window goes on the inside; if it opens inward, the storm is attached on the outside. Either way the window cannot be opened while the storm is in place.

Double-hung windows always have their storms attached to the outside of the sashes and they can always be opened. You can have a simple wooden or metal frame that hangs from hooks screwed to the outside of the house. The storm is then removed in the summer and replaced by a screen that is fitted into its own frame. But that requires some hard, and often dangerous, work twice a year, particularly when you have to haul a heavy storm window up a tall ladder to hang it over a second-story window.

Two- and three-track storm and screen arrangements were devised a few years ago and are now in wide use around the country. Essentially, the track systems consist of an aluminum frame with separate tracks for one or two storm windows and a screen. With the two-track system, the storm window is removed from the inner track and stored during the summer. The screen may or may not be taken off during the winter. The three-track system has the upper window on the outside track, the lower window in the center track, and the screen in the inside track, but none of the sashes are larger than half the height of the window. With either system you can slide the sash up and down or remove it for cleaning with considerable ease, by pulling it down to the small notches in the bottom of the track and popping it free of the frame at that point.

WHAT TO LOOK FOR WHEN BUYING STORM WINDOWS

Ask about the gauge of the metal. The heavier the gauge of metal used, the more durable the storm window and its frame. You will pay more for a heavy-gauge metal storm, but if a light-gauge frame is so light that it bends under a slight touch, it won't last for more than a few years, and then you will have to buy new ones. Spend your money wisely from the start.

Buy a unit with either an anodized or baked enamel finish. Naked aluminum will oxidize and become pitted, as well as be difficult to operate, after a few years of facing up to the weather.

Examine the corner joints. Each corner should overlap, not come together in a miter. If you can see through the corners, you will not only have problems with air infiltration, but the inherent strength of the unit is poor and it will not last long.

Check the locks, catches and other hardware for strength and ease of use. Also look closely at the joints between the glass or screen and the frame. It should be a soft, tight seal.

Try out the sliding panels. If they are hard to work in the store, think what they will be like after a year of standing out in the weather.

Shop around a lot. Compare different brands and designs. You will be spending considerable money and should be paying for a product that will last long enough for you to get your money back in reduced heating and cooling costs.

You can save money if you install the storm or storm-and-screen-combination windows yourself. The combination units normally fit tightly inside the window frame and are held in place by screws. Your major concern is that the frame be absolutely square so that the panes can slide freely in their tracks. If you attach a frame and then

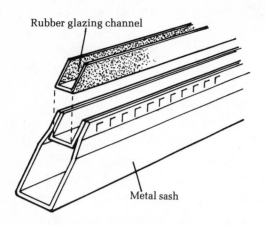

Remove the rubberized channel in the aluminum sash pieces before cutting the aluminum

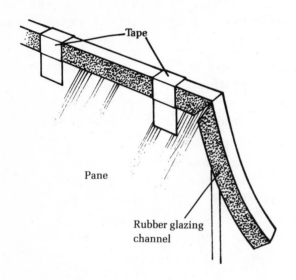

The sides of the rubberized channel are cut to form a 45° mitered corner

MAKING YOUR OWN STORM WINDOWS

You can save even more money by making your own storm windows. The components can all be purchased at most large building centers, hardware stores or lumber yards, either separately, or in kit form. But price the cost of the channels, corner locks, and other hardware against that of assembled units, and also compare the price of glass with that of clear acrylic plastic sheets. If the plastic is cheaper, it will do just as well as glass and probably will never break or need to be replaced. You can, of course, use wood for your frames, but to work with wood properly you should have a bench or radial arm saw and a router or dado blades, as well as the skill to use them. It is easier to work with aluminum framing.

How to Make Aluminum Storm Windows

1 • Measure the outside of the window frame at least three times.
2 • Remove the rubberized glazing channel inserted in the aluminum sash pieces and cut the aluminum with a hacksaw and miter box. Each corner must be mitered 45°.
3 • Install the corner lock hardware into the ends of the two side pieces. For accuracy in measuring, lock the two sides to one of the horizontal pieces.
4 • Measure the length and width of the sash from its outside edges and subtract 1/16″ from each dimension to allow for the frame width. Have the glass or plastic cut according to your measurements, or cut it yourself.

discover that it is out of line, check it with a square and then loosen some of the screws until it is square. If the gap between the frame and the unit becomes too wide, fill it with shims. After the frame is installed, caulking should be beaded around both the inside and outside edges.

5 • Beginning at one corner, insert the rubberized channel over the edge of the glass or plastic. Miter the first corner with a sharp razor blade. Attach the channel along the edge, holding it in place with tape, if necessary, until you reach the next corner. At the second corner, cut a right angle out of both sides of the channel, but do not slice through its back. When the channel is folded around the corner the two cuts should come together to form a 45° miter. Continue to bend the rubberized channel around the perimeter of the glass or plastic until you get back to the corner from where you started. Cut the last end at 45°.

6 • When the channel completely encases the pane, insert it into the two end pieces of its frame, then install the side frame pieces. Install the adhesive-backed weather stripping around the inside edge of the frame.

STORM DOORS

A solid wooden door, despite the fact that it is opened and closed all day long, will still conduct considerably less heat than a glass window. Doors also allow hot air to infiltrate into your home all summer long, which puts an added strain on your air conditioning. If the door happens to be of the large, sliding variety, the best you can do to prevent it from wasting away huge amounts of heat is to have it made of double-glazed thermal pane. It will cost more initially, but will pay for itself in surprisingly short order in reduced heating costs.

It might be argued that since doors are so

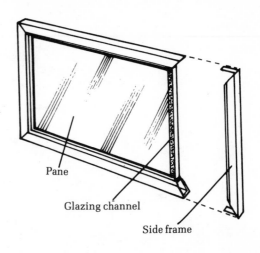

Pane

Glazing channel

Side frame

Install the glazing in its frame after the rubberized channel is in place around its edges

relatively energy-efficient you can get away with nothing more than some good weather stripping and lots of caulking. But a storm door can reduce outside noises, dirt and pollutants, and it also protects the primary door from weathering. If the door is really drafty, you can install a small removable vestibule around it, which will act the same way the air space between a storm window and its primary window acts, and then hang a storm door on the vestibule.

If the vestibule is not of interest, there are alternatives to standard wooden doors. A carved outside door costs in the neighborhood of $150 and up. For around that same price, you can purchase metal doors built around a core of insulation that gives them an R-value of about 15. Also available are doors made of rugged polypropylene, which have glass and screen panels that slide up and down from their bottom half and will retain 45% more heat than a standard solid wood door. The doors of man-made materials tend to be marketed complete with

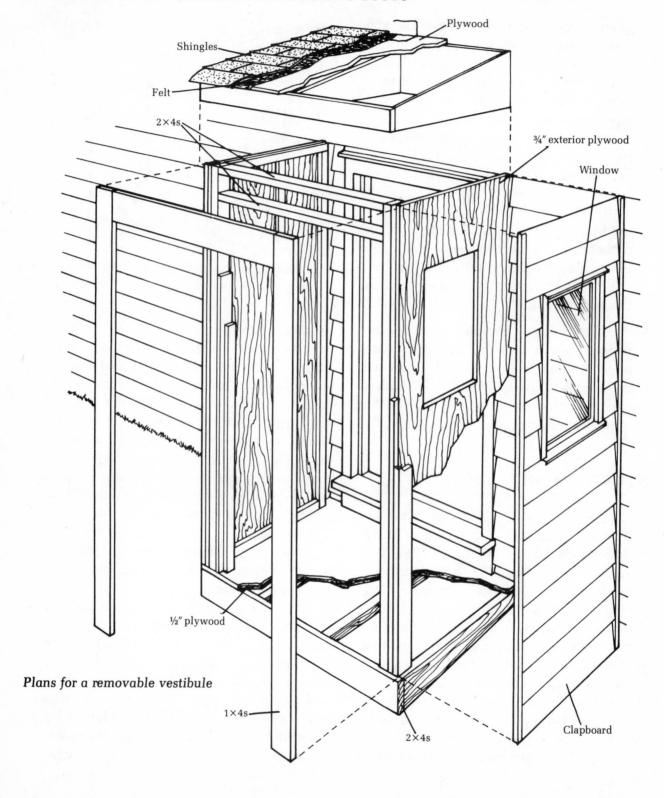

Shingles

Plywood

Felt

2×4s

¾" exterior plywood

Window

½" plywood

Plans for a removable vestibule

1×4s

2×4s

Clapboard

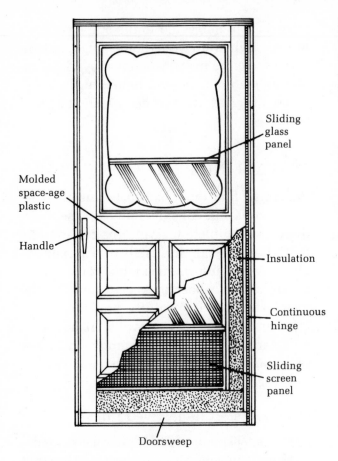

Sliding glass panel

Molded space-age plastic

Handle

Insulation

Continuous hinge

Sliding screen panel

Doorsweep

Exterior doors of man-made materials are extremely energy-conscious, and are competitively priced with standard wooden doors

frames, weather stripping, sweeps and other energy-conscious accoutrements that make them worth considering any time you decide to replace an exterior door on your house.

If you are content with the primary doors you have, it still makes sense to add a storm door of some sort. But don't buy cheaply. A storm door made of thin-gauge aluminum will quickly sag, warp and bind. As soon as the door stops fitting snugly in its frame there will be so much air infiltration you might as well not have any door at all. To add to the misery, the glass will fall out the first time a strong wind gets at it.

When you buy a storm door, make certain that it is made of heavy-gauge metal and is equipped with either safety glass or an acrylic plastic that will not shatter. It should also have an automatic closer that can be regulated to control how quickly the door locks shut, and can be adjusted to hold the door open.

A storm door is installed like any door, with its hinges attached to the door frame. Installation is a matter of positioning the door correctly, drilling holes for the hinge

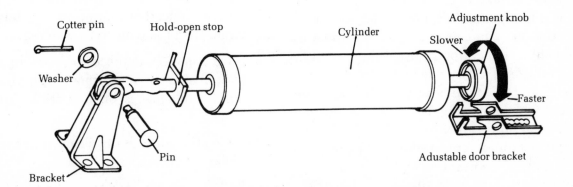

Cotter pin

Hold-open stop

Cylinder

Adjustment knob

Slower

Washer

Faster

Pin

Bracket

Adustable door bracket

Components of a door closer. One should be installed on every storm door

screws, and driving the screws home. Begin by installing one screw in the top hinge, then aligning the door in the frame and inserting one screw in the bottom hinge. Test the door to be sure it opens and closes properly. If it does, insert the remaining hinge screws. If it does not hang properly, remove either the top or bottom hinge screws and realign the door. Connect the door closer to the frame only when you have hung the door and it is working properly.

REFLECTIVE FILM

One of the tricks of modern technology is a group of products known as reflective film. Solar film, as it is sometimes called, has been used on windows in large office buildings for some time, and from outside the building it looks like the glass is mirrored, bronze, gold or silver. The hue and substance of the film is actually reflecting the sun's heat away from the building during the summer. During the winter, it reflects whatever heat is inside the building back into the building.

The film comes in different degrees of reflectibility. You can buy what is known as 50% film, which looks just a little smokey but still does a very adequate job of reflecting heat. Or 65% film, which offers an even better job of reflecting, but may look like you have put up mirrors instead of glass. The bronze-colored films reflect between 80% and 85% of the heat that strikes them.

In addition to reflecting heat, the reflective films provide some added benefits: reducing the fading in draperies, carpets and other fabrics subjected to the sun's rays; allowing light into your home while giving you almost the privacy of one-way glass; reducing glare inside your home; turning any piece of glass into a nearly shatterproof pane.

INSTALLING REFLECTIVE FILM

Reflective film once had to be installed by professionals at considerable expense. Now there are complete tool-and-instruction kits that allow anyone to install it. The kits include everything you need, but you can also get by with a rubber squeegee, spray bottle, metal straight edge, utility knife, masking tape and some paper towels. Installation will vary slightly among brands, so follow the manufacturer's instructions carefully. But in general, the steps are these:

How to Install Reflective Film

1 • Mix a teaspoon of liquid dishwashing detergent in one pint of water and pour it into a spray bottle.

2 • Spray the window with the detergent solution and wipe it dry with paper towels.

3 • Cut the reflective film approximately ½" larger than the pane to be covered.

4 • Stick some short pieces of masking tape to both sides of one corner of the film. Do not let the pieces of tape stick to each other.

5 • Spray the glass with clean water, completely covering it.

6 • Pull at your masking tape tabs so that the backing peels away from the film. As you work, spray water on the sticky side of

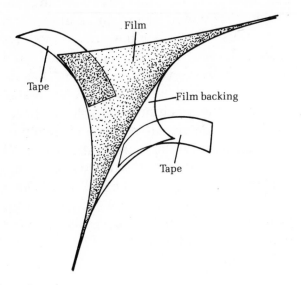

Attach masking tape tabs to both sides of the solar film to facilitate peeling off the backing

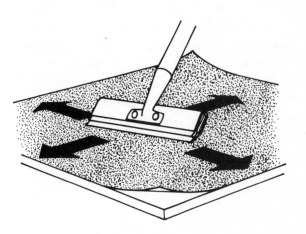

Squeegee the film until it adheres to the glass

the film. The entire sticky surface of the film must be wet.

7 • Position the sticky side of the film against the glass. The water will allow it to slide around quite easily, so you can position it with no trouble. Spray the outside of the film with water when you have the film in place.

8 • Squeegee the film, working from the center out to the edges. You can wrap the blade of your squeegee with paper towels to help soak up water, and don't worry about bubbles or any haze that may settle over the glass. Both will disappear in a few days. Also do not squeegee the edges until they have been trimmed.

9 • When most of the film has been squeegeed, trim its overage to within 1/16" of its edges, then squeegee the edges.

The film will need a few days to cure and it should never be cleaned with any cleanser containing ammonia. If the edges come loose, paint them with clear nail polish and press them back in place.

ADDITIONAL PROTECTION FOR WINDOWS AND DOORS

There are other materials that will help to keep out the summer heat and winter chill, thereby saving still more energy costs. These include blinds, shutters, shades, drapes and awnings.

Either blinds, shutters or shades can reduce the heat coming through your windows by as much as 50% during the summer, and reduce heat loss in cold months by as much as 25%, so they represent a considerable financial and energy savings. They should always be considered as a viable addition to your windows.

Draperies have come a long, long way toward saving energy. They have come so far, in fact, that you can now find several kinds of thermal-backed drapery fabric and drapery liners that will effectively keep heat inside your home during the winter and prevent it from getting through your windows in the hot summer months. Whenever you consider replacing the draperies you now have, think first about the thermal fabrics.

Awnings may represent the biggest savings of all: as much as a 75% reduction in heat gain from the sun's rays. Awnings installed over windows and doors in east and west walls will diffuse the sun's direct rays through most of a summer day. If you install them only over every window that has an air conditioner, they will help considerably to reduce the machine's work load on hot days. You can install permanent awnings, but during the winter these continue to shed sunlight away from the windows, at a time when the sun's rays could be helping to heat the house. It is more effective if you put up awnings that can be raised and lowered, or removed during the winter.

5 HEATING systems

Your home may have a warm-air central heating system, which heats the rooms by blowing heated air from the furnace through ducts that end at discrete gratings placed in the walls of each room. Or you may have a steam or a hot water system, which send, respectively, steam or near-boiling water through pipes to radiators in each room. In each of the three ways we centrally heat our homes, the air or water is heated by a furnace located in the basement or in some other out-of-the-way place. The furnace can be fueled by oil, natural gas, coal or electricity and is controlled by a thermostat which automatically turns it on whenever the temperature in the house falls below a preset temperature. Alternatively, whenever the house is warmer than the preset temperature, the thermostat prevents the furnace from turning on.

THE CARE AND MAINTENANCE OF HEATING SYSTEMS

Every heating system needs attention from time to time; in return, its improved efficiency will require less fuel and, therefore, help to reduce the cost of heating your home. Here are some general maintenance rules to observe:

1 • Keep all the filters, blowers, ducts, vents and the thermostat clean.

2 • Be sure the thermostat is not positioned in a draft, on a cold outside wall or near any heat-producing appliance (such as the stove, television or a lamp) which might cause it to give a false reading of the room's temperature.

3 • Unless it is the dead of winter, when your pipes might freeze, turn off the fur-

nace whenever you plan to be away from home for more than a day or two. If it is the middle of winter, at least reduce the thermostat setting to 60°F. When you come home, don't jack the temperature up to 85°F to warm the house up faster. It won't happen. Anytime you start the furnace it will generate heat throughout the house, and the temperature will reach your normal setting in the same amount of time as usual. But, by setting the thermostat higher than normal you may forget to turn it down again and waste fuel by overheating the house.

4 • Lower the temperature when you are cooking or have more than the usual number of people in the house. Both food preparation and people generate enough heat to offset turning the thermostat down.

5 • Whenever you have the heat on, but particularly at night, pull drapes over all windows and glass doors. Glass is a conductor of heat. Also keep the damper to the fireplace closed whenever there is no fire. An open chimney is as efficient at sucking hot air out of the house as an open door.

6 • Shut off the radiators or vents and close the doors to any rooms that are not regularly used.

7 • Do not block radiators or vents with draperies, furniture or covers. The heat will be unable to circulate properly.

8 • Attach automatic closers to all exterior doors, especially if there are children living in the house.

9 • Let hot bathwater cool before draining it

out of the tub. Both the heat and moisture from warm water can help to keep the house comfortable.

10 • Maintain the proper humidity (you may have to install a humidifier). It has been proven that if the air is too dry, it can feel uncomfortable even if it is at a comfortable temperature setting.

HOW HEATING SYSTEMS WORK

Each heating system has its advantages and drawbacks. You can argue for and against your particular system forever, but the best way to cohabit with it is to understand how it operates and try to maintain it at peak efficiency. A hot water system is more expensive to install, but will operate almost trouble-free for years. Warm air systems are inexpensive to install and also offer the versatility of letting you add both a humidifier and central air conditioning. No one installs steam heat systems anymore, so they are found only in older homes and are expensive to replace. But the fact is that all central heating systems can—and do—operate for decades without any great demand on your time, energy or wallet.

WARM AIR SYSTEMS

A forced air furnace can be fueled by gas, oil or powered by electricity. Cold air is drawn into the furnace by a fan attached to the return (cold) side of the furnace. The cold air is drawn through a filter, then warmed by the furnace and allowed to rise through a duct system that carries it to every room in the house. If there is a humidifier, it is con-

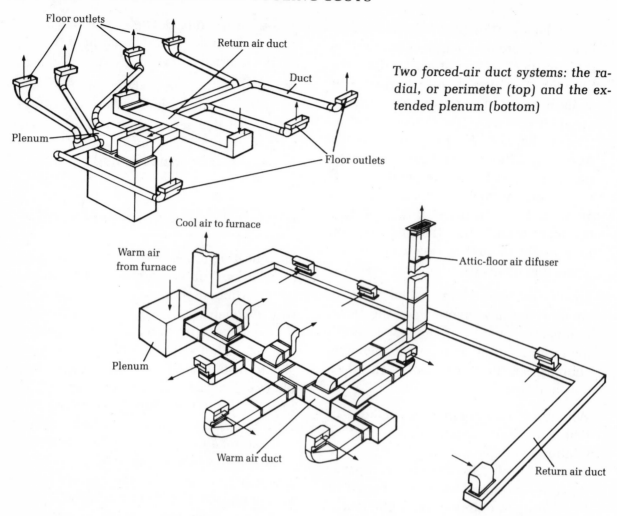

Two forced-air duct systems: the radial, or perimeter (top) and the extended plenum (bottom)

nected to the warm air side of the furnace so that it can add moisture to the air entering the rooms.

The registers in each room heated by a warm air system are designed to diffuse the heated air evenly, without leaving any cold drafts in the room. They may be located near the ceiling or just above the baseboards, but should not be too close to the return registers. The return register is supposed to funnel cool air back to the furnace, where it is warmed, so they should always be placed near the floor.

The key to any forced air system is its ducts, which can be arranged in either the *radial* (perimeter) configuration, or in a form known as the *extended plenum*.

Radial System

The radial system reaches from the hot side of the furnace directly to the outside walls and circles the house with at least one register in the outside wall of each room. There is also a return air register (grate) in each room, which leads to the return air ducts, which go directly back to the cold side of the furnace. The return registers are usually placed on an outer wall, if the room has one.

Extended Plenum System

The extended plenum system uses larger, rectangular ducts that reach out in all directions from the top of the furnace to each room in the building. The return air ducts are slightly smaller; they come from every room back to the cold air side of the furnace.

Maintaining a Forced Air System

Keep the registers free of dust and dirt by vacuuming them whenever the room they are in is cleaned. Also, check the ducts from time to time for air leaks, by running your hands along their surfaces and around their joints. If you find any leaks, wrap them with duct tape.

Most homeowners rely on an annual service inspection to keep their furnaces in running order. By and large, that inspection will prevent any major catastrophes from occurring; a good inspector will detect when parts of the furnace are on the verge of breaking down. A part of that annual inspection is to clean or replace the filters in the furnace. However, the filters in a forced air heating system ought to be changed more than once a year and, in fact, some filters should be changed about once a month.

The filter is normally positioned between the blower (fan) and the furnace vent in the return air ducts. To replace or clean the filter, open the access panel to the furnace and remove the old filter, which resides in either a frame or some sort of removable device that is simply pulled out of its cavity. Filters that are not supposed to be replaced can be cleaned by vacuuming, by wiping them with a cloth, or according to the cleaning instructions pasted on the side of the heating unit. When the filter is clean, or the replacement filter is installed in its frame or whatever holds it, slide it back into its cavity. The size of the replacement filter is normally printed on the frame along with an arrow which must point in the direction of the air flow when the filter is installed. In most cases, the arrow will be aimed at the furnace when you install the filter.

Besides the filter in a forced air system, there are other places that can accumulate dirt, which in turn inhibits the proper flow of air through the ducts. The most critical of these areas is the blower itself, particularly if you have a squirrel-cage type of unit. When the openings in the cage are clogged, air cannot move through the unit efficiently, with the result that the blower will have to work harder and longer, costing you more for both electricity and furnace fuel.

Before you even touch the blower, disconnect the electricity that powers it by removing the proper fuse or shutting off the appropriate circuit breaker. Then follow this procedure:

1 • Remove the filter access panel and undo the retaining screws that hold the fan in position.

2 • Pull the fan off the frame. If the power cord is too short to get the fan away from the furnace, disconnect it and label each of the wires (tie a tag to them) so that you will know exactly how they go back together.

3 • Every blade on the fan must be cleaned with a brush. You may find the biggest tool you can use to get at the blades is an

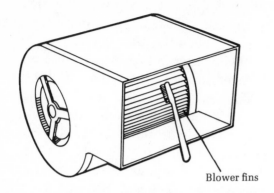

Blower fins

The blower can be cleaned with an old toothbrush

old toothbrush. Clean the blades and any area around the fan that you can reach. Then vacuum the entire unit.

4 • Reconnect the power cord (if you had to disassemble it) and put the fan back the way you found it. Tighten the retaining screws, close the access door, and turn on the electricity.

The blower motor may have oiler cups on its housing or it may be an entirely sealed unit, but both versions should be lubricated

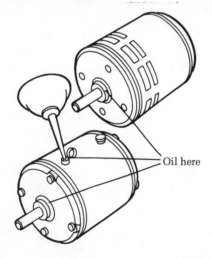

Oil here

Lubricate the blower motor at least once a year

at the start of each heating season. If the motor has oiler cups, squirt 5 to 10 drops of light machine oil into each cup. If the unit is sealed, put 5 to 10 drops of oil on the pad that surrounds the shaft as it enters the motor.

If the blower is connected to its motor by a drive belt, spray the belt with *spray belt dressing* any time you hear it squeak; spray it at least once a year whether it makes any noise or not.

A half an hour or so each year devoted to cleaning and oiling the blower and its motor will not only reduce your utility bills by increased operating efficiency, but will prolong the lives of the parts and avoid costly repair bills.

HOT WATER SYSTEMS

Hot water heating can be accomplished with a *gravity,* *hydronic* or *radiant* heating system. Water is heated in the furnace boiler and pumped through a system of pipes to each room. As the water cools it returns to the boiler to be heated again. Hot water systems utilize not only a house thermostat, which turns on the circulating pump whenever it calls for more heat, but also a second thermostat on the boiler itself, which turns on the boiler any time the temperature of the water falls below a preset temperature.

Gravity System

When water is heated, it loses weight and tends to rise in the same manner that warm air rises. In the instance of a gravity hot water heating system, the heated water rises out of the boiler and through the pipes to the radiators. It is instantly replaced in the

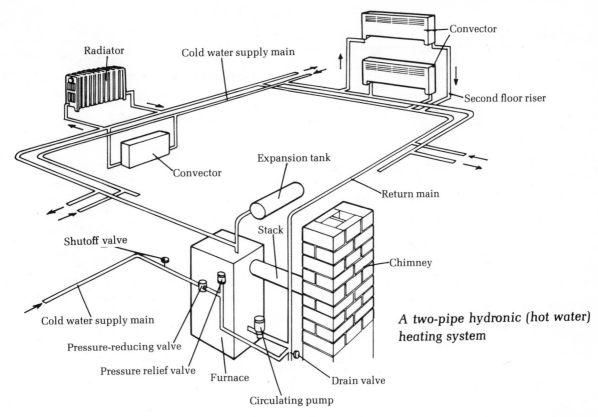

A two-pipe hydronic (hot water) heating system

boiler by colder water returning from the radiators via a set of return pipes. The pipes in a gravity system are sealed and full of water that is constantly expanding and contracting as it is heated to about 180°F then cools to around 140°F. There is also an expansion tank, which contains only air, somewhere near the boiler. As the heated water expands, some of it enters the tank and compresses the air, which then puts pressure on the water, forcing it to continue moving through the system.

Hydronic System

Also known as a forced hot water system, a hydronic system is really nothing more than a gravity system with a circulating pump.

The pump is attached to the return pipes to help drive the water through the system, and is controlled by the house thermostat so that it goes on whenever the thermostat calls for heat. It also has a control valve that stops the flow of water anytime it is not running so that the heated water will not rise through the pipes and overheat the house.

Hydronic systems can heat the water that comes out of your faucets, which means there is no need for a separate hot water heater. On the other hand, anytime the house water falls below its predetermined temperature, the furnace turns on, whether during a blizzard or a heat wave. The water is heated by passing it from the cold water main through a set of coils in the back of the furnace.

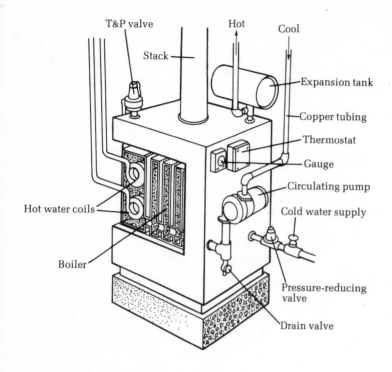

A hot water heating system furnace

Series-loop pipe arrangement for a hot water heating system

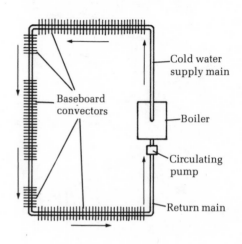

Two-pipe arrangement • Originally, hot water heating systems had one set of pipes that brought the heated water to each radiator, and another set of pipes for the cooled water to return to the furnace. Many older houses still have their original two-pipe system.

Series-loop arrangement • Newer, smaller houses save the cost of piping by using the series loop arrangement, which is simply a single pipe that goes from the boiler through each radiator in the house and back again. So long as none of the radiators are closed off, the series loop is fine. But if you shut off one radiator, you have closed down the entire system. In large houses, which might have different sections requiring different temperatures, more than one loop must be

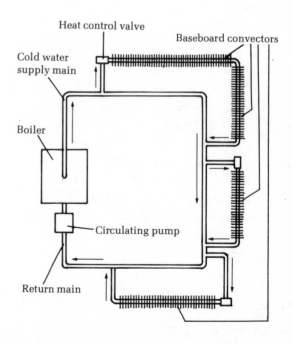

Single-pipe hot water heating system

installed or some other pipe arrangement used.

Single-pipe arrangement • This is a pipe that goes from the boiler to every radiator, and back to the boiler, as if it were a series loop. The important difference is that each radiator is connected to the pipe with a short branch line and control valve so that any radiator can be turned off without closing down the entire pipeline.

Radiant Heating Systems

Another version of the hot water system is to bring the heated water into a series of pipes buried in the walls, ceilings, or floors. Heat from the pipes radiates through the surface material (plaster, concrete, etc.) to warm the room.

The pipes must be made of copper and are known as coils, and the system is considered the most comfortable heating system possible. It is also the most expensive to install. Not only is the cost of copper piping extremely high, but installing the panels of pipe is time-consuming and expensive.

Maintaining Hot Water Systems

Aside from cleaning the radiators or convectors, little can or needs to be done to maintain a hot water heating system. The furnace and boiler should be cleaned at least once a year. If there is a pump, it should be cleaned and oiled at the same time.

Radiant heating system

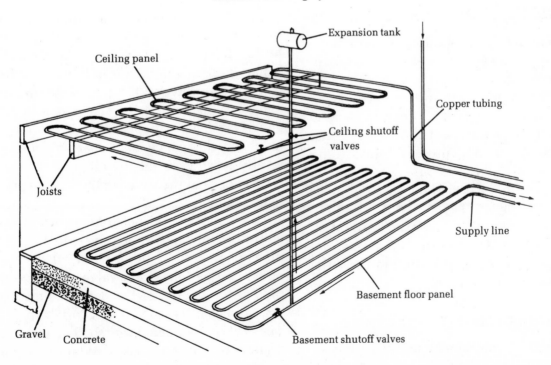

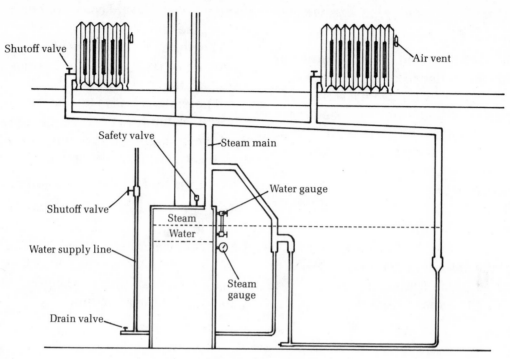

Steam heat pipe arrangement

STEAM HEATING SYSTEM

Essentially, a steam heating system functions the same as a hot water system, except that the water in the boiler is heated to 212°F so that it boils and turns to steam. The steam rises through a series of 2″ pipes that lead to each of the radiators. As it cools in the radiators, the steam condenses back into water and runs down the same pipe it came up, to be reheated in the boiler. Because the pipes are 2″ in diameter, they are large enough to handle the returning water at the bottom and the rising steam at the top. Some systems use separate return pipes to carry the water back to the furnace.

Maintaining a Steam Heating System

Water in the furnace-boiler must be replaced periodically, so it should be checked at least once a week. There is a glass tube attached to the side of the furnace that should always be half filled with clean water. Anytime the water level in the gauge is low, open the water intake valve on the side of the furnace and leave it open until the water in the gauge reaches its proper level. If you forget to check the gauge and hear steam constantly hissing out of the radiators it is an indication that the water in the boiler needs to be refilled.

The water should be drained out of the entire system at least once a year to clear the pipes of rust and sludge. Shut off the furnace and open the drain valve in the steam pipe at the bottom of the boiler. The valve usually has a threaded spout to which you can attach a garden hose so that you can drain into a pail or basin. When no more water comes out of the valve, close it and open the shutoff valve. Fill the boiler until water reaches the proper level in the gauge tube.

Steam boilers have a pressure gauge positioned on top of the boiler. The steam pressure must read between 2 and 10 pounds per square inch (psi) so when the pressure reaches 12 psi it is too high; at 15 psi a safety relief valve will release the excess steam. If it does not blow off steam, the boiler is likely to explode, so be certain the valve is functional. Test the pressure relief valve every three or four months by manually lifting its lever. If no steam comes out of the valve, replace it.

If the knocking and rattling that is normal for any steam system grows too loud or goes on incessantly, there is water trapped somewhere in the pipes which is blocking the steam. The pipes in a steam system all slope toward the furnace so that the water can always drain back to the boiler. Sometimes a pipe will break loose from its fastenings and sag, forming a low area where water can collect and block the steam from rising past it. If the knocking becomes unbearable, trace each of the pipes from their radiators to the

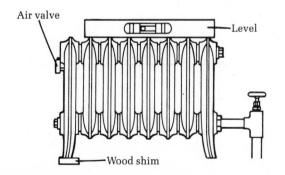

Every steam radiator must tilt toward the pipe that serves it. The radiator can be shimmed at one end

furnace and refasten any pipe that is sagging. Should you decide the pounding is coming from one of the radiators, slide small pieces of wood under the radiator legs farthest from the pipe so that the radiator is pitched more severely, allowing water to run out of it, back to the boiler.

RADIATORS

Hot water and steam heating systems employ cast-iron radiators, convectors or baseboard radiators. The *cast-iron radiators* work on the principles of convection and radiation. When the radiator is hot, cool air touching it is heated and rises toward the ceiling; any object near the heated radiator is warmed by waves of heat radiating from the metal. How much heat the radiator gives off is proportionate to how much metal is exposed, although at least 30% of the heat in any radiator is radiated from the unit. A long, low radiator will give off more heat than a short, high one, if only because the longer unit has a larger volume of heat to shed. For the same reason, a covered radiator will be less efficient than a freestanding unit.

If you have to cover your radiator, make certain there is ample space between it and the cover to allow air to circulate. It is a good idea to place a sheet of reflective metal behind the unit to assist in radiating the heat.

Convectors are relatively new cabinets that contain a tubular metal heating element surrounded by metal fins. Room air enters through a grille at the bottom of the unit, is warmed by the tubular heating element, which is full of steam or hot water, and then rises past the fins and out the grille

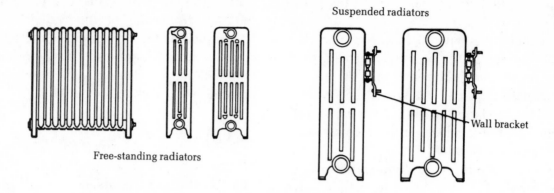

Cast-iron radiators are used for both water and
steam heating

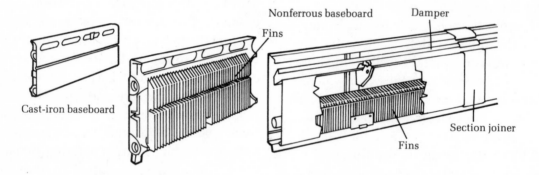

Cast-iron baseboard heaters can be used with either steam or hot water systems (left). Nonferrous
baseboard radiators (right) can be used only with a hot water system

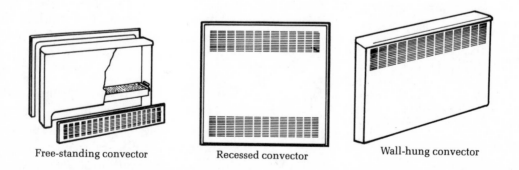

Convectors are also used for steam and hot water
heating

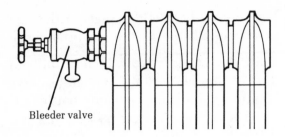

Hot water valves can be bled by opening the valve knob until water flows freely from the valve

Bleeder valve

at the top of the convector. Because the fins present a number of heated surfaces, the convectors are more efficient than most cast-iron radiators.

The most efficient radiator of all for delivering warm heat at floor level is the *baseboard radiator*. These are narrow (4″) and come in lengths of from 12″ to 8′. Inside the cast-iron housing is a pipe or pipes surrounded by tiny fins so that air moving under the radiator and up past the fins is radiated and convected in the same manner as with convectors.

Radiator Valves

All radiators have an air valve which releases air trapped inside the radiator with a hiss as the unit fills with steam or hot water. A valve can become clogged with dirt, and its tiny air hole is easily filled if the valve is painted. Valves can also leak, and occasionally water becomes trapped inside them.

Hot water valves • If the valves attached to hot water radiators trap any water, they are drained by "bleeding." Simply rotate the vent valve knob on the radiator until hot water squirts out of the valve, then shut off the vent valve.

Steam valves • Should the valve on a steam radiator become clogged, shut off the radiator by turning the valve at its base counterclockwise. Unscrew the air valve from the radiator and remove it. An air valve can be cleaned by soaking it in vinegar overnight, then flushing it with water. Air valves should be replaced if the radiator does not seem to heat up properly. Use any of the different kinds of replacement valves sold at most hardware stores.

THERMOSTATS

The heart and soul of any central heating system is the thermostat that controls it. Thermostats are extremely sensitive instruments designed to react whenever the temperature in your home is not exactly at the level you have set it for. They can wear out, but all kinds of replacements, either manually or automatically controlled, are available. The automatic units incorporate a timer and at least two temperature setting levers, so that the unit will hold the heat in

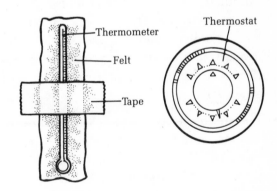

A thermometer used for testing the thermostat of a heating system must be insulated from the wall temperature so that it can give an accurate reading of the surrounding air

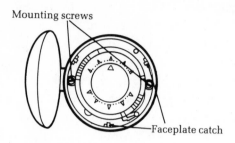

The thermostat cover can usually be pried away from the base

your home at different temperatures during the day or night.

If you have set your thermostat at a particular setting and the temperature in your house is either warmer or cooler than that setting, the thermostat may be incorrectly calibrated. If the unit can be recalibrated, you can do it yourself by following the instructions in the owner's manual. Otherwise the thermostat must be replaced.

How to Check a Thermostat

1•Tape or hang a tube thermometer on the wall next to the thermostat, placing a piece of felt or padding between it and the wall so that the thermometer is not influenced by the temperature of the wall or a draft.

2•Give the thermometer 15 to 20 minutes to stabilize and then compare it with the thermostat thermometer. There should be no more than a one-degree differential. If there is a difference, dismantle the thermostat and clean it.

How to Clean a Thermostat

1•The glass (or plastic) faceplate on a ther-

mostat usually is held in place with snaps or a friction catch so it can be pried loose.

2•With the cover removed, blow off any dust in the thermostat. You can use your breath or a plastic squeeze bottle, but never a vacuum cleaner because the suction is too powerful.

3•If you find any contact points without further dismantling the unit, clean them by sliding a piece of paper between them, or spray them with contact cleaner.

4•If you discover a mercury vial inside the unit, make certain it is absolutely level. If it is off level, loosen the mounting screws in the rim of the thermostat and adjust the unit until the mercury reads level.

How to Replace a Thermostat

If cleaning and leveling the thermostat do not correct its temperature reading, and if the unit cannot be recalibrated, replace it with a unit that has the same voltage as the old unit and is compatible with the heating system in your house. Thermostats normally use so low a voltage that you can wire them into your house system without turning off the electricity, but don't take any chances; turn off the current by tripping the circuit breaker or removing the proper fuse. Then follow this procedure:

1•Remove the faceplate on the thermostat and undo its mounting screws. Pull the unit away from the wall.

2•Disconnect the low voltage wires leading from the back of the thermostat by loosening their terminal screws. Be careful not to let the wires fall into the wall, out of reach.

3• Scrape the ends of the wires with a knife until they are clean and shiny, then wrap them around the terminal screws on the back of the new thermostat and tighten the screws.

4• Push the excess wires into the wall. If the hole in the wall is larger than the new thermostat, tape the opening so that cold air in the wall cannot escape into the room.

5• Position the new thermostat on the wall and hold it there with one of the mounting screws. Rotate the unit until the mercury in the tube shows that it is level, then attach the second mounting screw. The thermostat must be level before it can operate properly.

6• Put the faceplate on the thermostat and rotate the dial to make certain the unit is turning your heating system on and off. Set the thermostat to the desired temperature or temperatures.

FURNACES

Furnaces can be oil-fired, fueled by natural gas or powered by electricity. There is no particular advantage in using one fuel over another so far as the efficiency of the furnace is concerned, but some parts of the country almost exclusively use natural gas or oil or electricity, depending on its most available resource. Generally, natural gas is considered to burn cleaner than oil, and electricity is deemed to be the most expensive of the three energy sources. It is questionable whether natural gas is any cheaper than oil in most parts of the United States. If you are selecting a heating system for a new con-

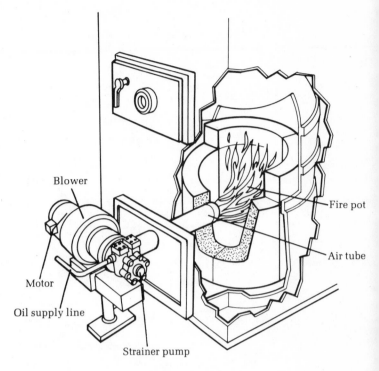

An oil-fired furnace

struction or as a replacement, most likely your choice will be based as much on fuel availability as its cost. That cost, of course, should be estimated not on today's prices, but on predicted price rises in the near and distant future.

OIL FURNACES

Oil-fired furnaces are used to operate water, steam, or forced air heating systems. Given proper annual maintenance, an oil furnace will give years of trouble-free service. The furnace typically consists of a high-pressure spray and a blower fan which sends a fine mist of oil into the combustion chamber where it is ignited by an electric spark. As long as the oil mist is blown into the chamber, the oil will continue burning. It is the

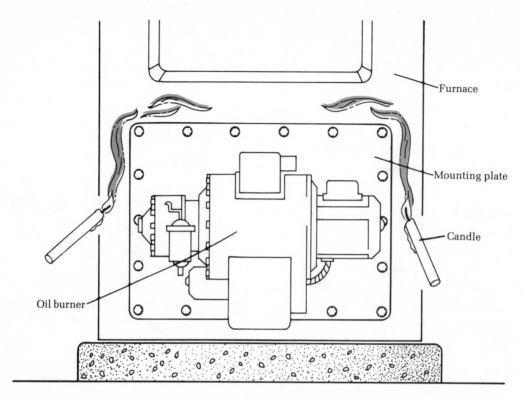

Move a lighted candle around the edge of the mounting plate. If the flame is drawn inward, the seal is poor and the gasket should be replaced

oil burner that must be attended most of the time.

Maintaining Oil Burners

Most electric motors require some lubrication at one time or another. Check the owner's manual for your oil burner to see when and where the fan motor should be lubricated.

Keep the blower (fan) clean. It should be scrubbed with a small brush at least once a year, or more often if the owner's manual recommends it. Always shut off the electricity before you do any work on the fan blades, or on any other part of the motor.

The stack is a large tin pipe, attached in sections, that goes from the furnace to the chimney, and gathers considerable amounts of soot during the course of a heating season. You can pull the sections of the stack apart, then bang them against a newspaper-covered floor to empty them. Be sure to connect them securely when you are reassembling the pieces.

There is a mounting plate between the burner and the combustion chamber which should be tightly sealed at all times. Light a candle and pass it slowly around the rim of the plate while the burner is running to check for air leaks. If the flame is sucked toward the plate, the seal is not airtight. Try tightening the bolts on the plate. If that fails,

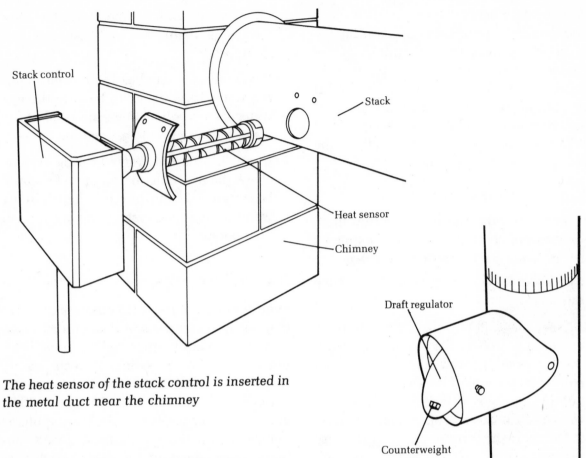

The heat sensor of the stack control is inserted in the metal duct near the chimney

The draft regulator is a revolving piece of metal with a counterweight attached

shut off the furnace, unbolt the plate and replace its gasket.

Inspect and clean the stack control every two months during the heating season. The stack control is a small box held to the side of the stack with retaining screws, and is usually found near the stack's entry to the chimney. Shut off the electricity to the furnace, undo the mounting screws, and pull the control away from the stack. A heat sensor, attached to the back of the box, normally resides inside the stack. There should be a light coating of soot on the sensor; clean it off and replace the control. If the sensor is heavily covered with soot, you are not getting adequate combustion in the furnace.

Call your serviceman for a complete overhaul of the furnace.

There is also a draft regulator on the stack, which is nothing more than a revolving metal plate that has a small counterweight at its top. When the furnace is running, the regulator should be slightly open; when the furnace stops, the regulator will close. If the regulator does not open and close properly, rotate the counterweight slightly until it functions.

The owner's manual for your oil furnace gives detailed information concerning the correct adjustments for such things as the air tube shutter, spark gap and other components that might have to be adjusted from time to time. Read the manual and check all of the areas suggested in it. Have a qualified serviceman inspect, clean and adjust the furnace before each heating season.

When an Oil Furnace Doesn't Run

Anytime your oil furnace fails to run, there are some immediate checks and adjustments you can make before calling a serviceman.

1 • Check the fuel gauge to be sure you have oil.

2 • Be sure the switch is on. If it is, and there is a reset button on the burner, push it. If the burner does not start, or if it starts and stops immediately, check the fuse or circuit-breaker box. If the fuse is blown or the circuit breaker is off, replace or reset it and start the furnace again. If the fuse blows or the circuit breaker goes off again, there is probably a short somewhere in the system. If you have electrical experience, locate the short and repair it. Otherwise, call a repairman.

3 • If the furnace keeps going out or is sporadic in its operation, it may have a dirty oil filter. Check your owner's manual for information on locating and removing the oil filter and cleaning it in kerosene. Kerosene is so inflammable that most stores are forbidden by law to sell it. When you find some, use it with extreme caution.

4 • Move the thermostat to a higher temperature. If the furnace switches on, but goes off at some lower temperature, the thermostat is malfunctioning. Replace the thermostat.

5 • If you have a steam heating system, check the water level gauge to be sure there is water in the boiler. Keep the level of the water halfway up the gauge tube.

6 • If you have a hot water heating system, check the recirculating pump to be sure it is working. If the pump is not moving water through the system, the burner will not operate.

GAS FURNACES

Even though natural gas is more expensive than oil in many parts of the country, a gas burner is generally preferred for home heating because not only does it costs less to install than an oil furnace, but it has fewer maintenance and repair problems. Besides, gas burns cleaner than oil. The problems that do arise with gas are almost always centered around the pilot, the electrical connections, or the thermocouple.

Maintaining a Gas Furnace

There are filters in gas furnaces that should be cleaned periodically. Check the owner's manual for specific information as to their placement and cleaning.

If you smell gas, or think there may be a gas leak, don't be a hero. Close off every gas valve you can find and call the gas company. Gas is extremely dangerous; let qualified professionals deal with it.

If a Gas Furnace Stops Working

There are a few simple repairs you can make

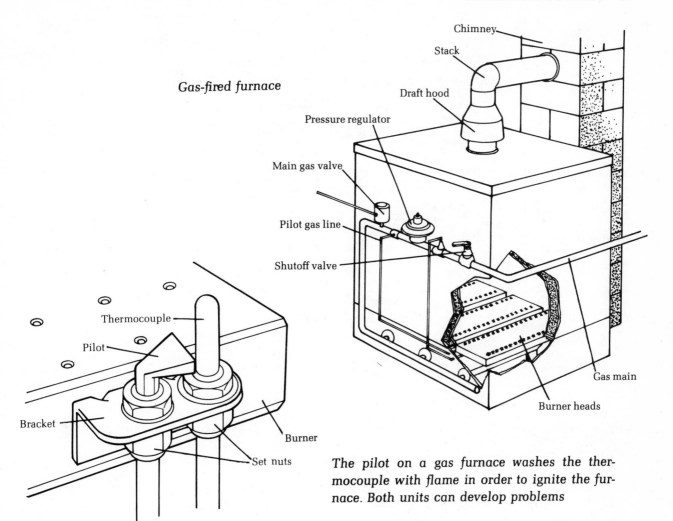

Gas-fired furnace

The pilot on a gas furnace washes the thermocouple with flame in order to ignite the furnace. Both units can develop problems

to a gas furnace if it stops working. If none of these get the system going again, call a repairman.

1· The pilot light can sometimes go out for no apparent reason, so always check the pilot first. Inspect the tiny orifice in it for clogging. You can use a needle to clean out the hole.

2· If the pilot light goes out repeatedly, the problem may be with the thermocouple, located opposite the pilot. The pilot flame must continually wash the thermocouple

in order to make the furnace operate. Check the owner's manual for information on the proper color and height of the pilot flame, and how to adjust the pilot to produce that flame (there is an adjustment screw on the pilot that will regulate the flame). The manual will also give you specific instructions for replacing the thermocouple.

3· If the furnace does not light, and the pilot and thermocouple seem to be in working order, the valve on the gas supply line

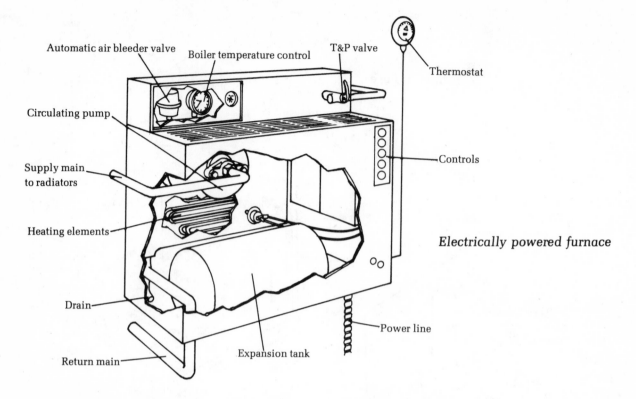

Electrically powered furnace

may be sticking or the thermostat may not be functioning. Hit the gas valve handle with your hand. It should move freely. If it is stuck, call the gas company to replace it. There is nothing hard about putting in a new gas valve, but the gas supply must be shut off and the proper seal has to be made around the valve to render the connection absolutely safe.

4•Check the thermostat by turning it up until the furnace fires. If only a higher setting starts the furnace, replace the thermostat.

5•Make certain that electrical current is reaching the furnace. Check for a blown fuse or tripped circuit breaker. Replace the fuse or reset the circuit breaker. If the electricity goes off a second time, the

entire electrical system must be searched to locate the short.

ELECTRICALLY POWERED FURNACES

The most expensive way you can heat any home is with electricity, whether you are using warm air ducts and an electrically energized furnace or individual baseboard units to heat each room. There are some advantages to the high cost of electrical heating, of course. Electricity is the cleanest of all ways you can heat your home. There is no need for a flue, which will carry off a percentage of your heat and also must be cleaned periodically, because no combustion ever takes place. Also there are no moving parts in an electric system except, perhaps, a blower fan or a circulating pump. Even with these, the repairs to an electrical system are always minimal.

Maintaining an Electric Warm-Air Furnace

1•Clean or replace the air filters at least once a month during the heating season.

2•Clean and lubricate the blower fan at least once a year.

3•Clean all vents and registers whenever their rooms are cleaned. You can vacuum and/or wipe the registers with a dust rag.

4•Occasionally check the duct work for loose joints or damage that might permit heat loss. Ducts can be sealed with duct tape.

When Electric Heating Stops

Before you call a professional, there a few basic checks you can safely make:

1•Check the main circuit breakers or fuses to be sure there is power reaching your heating system.

2•Shut off the power and open the furnace. Each heating element in the furnace has its own fuse or circuit breaker. Check to be sure none of the breakers are tripped or the fuses blown. These can be reset or replaced, but if they break the circuit to the heating element a second time, there is a short circuit somewhere that must be searched out and fixed by a qualified person. If you do not feel qualified, call a serviceman.

3•The last line of defense is to check the thermostat to be sure it is functioning properly. If the thermostat is functioning, call a serviceman to troubleshoot the wiring and heating elements.

If an Electric Baseboard Heater Fails

Some electric heating systems consist of

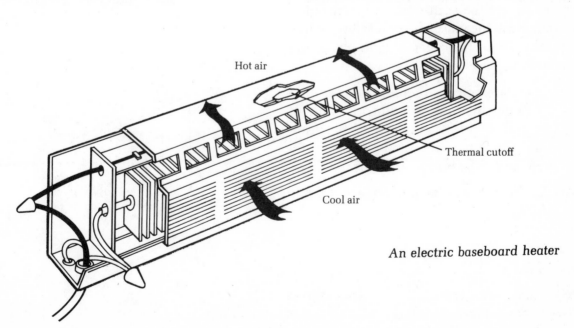

Hot air

Thermal cutoff

Cool air

An electric baseboard heater

baseboard heaters positioned along the walls of each room in the house. There is not much you can do to maintain baseboard heaters other than to clean them periodically. Most baseboard units are either wired directly into their branch circuits or plugged into a wall outlet, and simply send heat into the air by convection, although some units also contain a small electric blower. There are a few checks to make to determine the viability of a baseboard heater that fails to work:

1 • Check for a blown fuse or tripped circuit breaker. Replace the fuse or reset the circuit breaker. If either shuts off the electricity again, there is probably a short circuit in the heater, which must be located and repaired.

2 • Shut off the electricity and inspect the wiring inside the heater. Look for loose or dirty connections, which should be cleaned or tightened. If you have a multimeter, test the entire unit for continuity, including the thermal cutoff safety switch and the heating element. Replace any parts that are faulty.

3 • If there are any obstructions in or around the heater that might inhibit the flow of air, remove them. A piece of furniture could block the unit, or the bottom of a drape might be stopping the air flow.

4 • Check the thermostat and replace it if it is faulty.

FIREPLACES

Over the years, fireplaces and wood stoves have been relegated from the only source of central heating in the home to the position of a backup heating system. Fireplaces are fun to have and use on cold wintry days, and a blessing if your central heating system ever fails to operate. But they are awesome wasters of energy.

Most of the heat from an open fire goes straight up the chimney and out of the house. Furthermore, that heat can take as much as 20% of the warm air in your house with it every hour—air that you have spent considerable money to heat. Then, when the fire dies down the room air keeps on going up the flue unless it is securely closed off by a damper.

Heating experts and engineers have addressed themselves to the inherent problems of fireplaces and have developed a variety of designs and products that help reduce energy loss from the average fireplace. Some designs incorporate fans hidden in both sides of the fireplace to draw room air through chambers next to the fire-

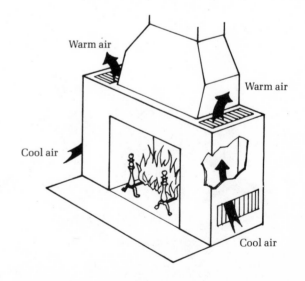

A fireplace design for reducing heat loss

place, where it is heated. The air is then sent back into the room or ducted to other parts of the house.

There are also convection duct units that fit into the fireplace. Cool air enters the bottom of the ducts, then rises as the fire heats it, and is ejected back into the room. Many heat-efficiency devices are beginning to proliferate the market and each should be considered carefully in terms of the kind of fireplace you have and the ways you can utilize the heat from an open fire.

Know that a fireplace is more of a heat waster than a provider. And if nothing else, be certain you have a good-sealing damper and that it is kept tightly closed whenever there is no fire in the fireplace. Even that is not truly efficient protection against heat loss unless you also have some sort of fireplace cover.

If your fireplace has no damper at all, one of the best ways of adding both a damper and a cover is to install a glass fireplace enclosure. These come in a variety of designs and incorporate a damper as well as doors that you can close around the fire to snuff it out. When open, the unit allows you to receive much of the fire's heat, and when closed you will lose almost none of the warmed air in your house.

THE QUESTION OF HUMIDITY

Studies made by both industry and the government have amply demonstrated that the proper amount of humidity in the house can make a considerable difference in comfort. Too little humidity, for example, will leave you feeling chilled even though the temper-

ature is a comfortable 70°F or so. In general, most people are comfortable when the relative humidity is between 30% and 50%. If the air in your home contained 100% moisture, it would be raining in your living room. Relative humidity means the percentage of moisture in the air compared to how much moisture the air could actually hold. Because warm air holds more moisture than cold, if you raise the temperature without increasing the amount of moisture, you will have a lower relative humidity. The result will be that you feel chilled, your nose and throat will be dry, and your wooden furniture will begin to come apart at its joints. On the other hand, if there is too much moisture in the air, that is if the relative humidity is more than 50%, you will have condensation on your windows, you will feel "clammy" and the joints in your furniture may swell so much the wood splits.

It should be noted that an air conditioner acts as a dehumidifier and removes moisture from the warm air of a hot summer's day. The appliance that acts in the reverse of an air conditioner is a humidifier, since it is designed to inject moisture into the air. There are some arguments against humidifiers. Proponents of the humidifier claim that they will reduce heating costs because you can maintain a lower temperature with a high relative humidity and feel very comfortable while saving the cost of fuel to raise the temperature in your house those extra few degrees. However, the humidifier is powered by electricity and that can cost as much as the fuel you save. While the cost factor may or may not be true, there is no question that the presence of a humidifier makes a considerable contribution to your

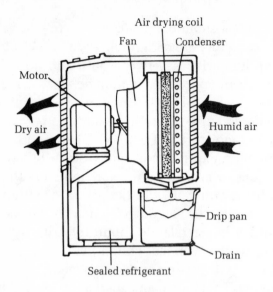

A portable dehumidifier

comfort, and that with it you can maintain your house temperature several degrees lower than without it.

TYPES OF HUMIDIFIERS

There are small tabletop-sized humidifiers, portable floor models and units that are designed to hook directly into your central forced-air heating system. All of these models are divided into two types: evaporator or atomizer. The *evaporator* versions have a reservoir of water positioned behind a heater and fan that blows warm air over the water. The warm air collects moisture particles from the water and carries them into the room air where they increase the relative humidity. The *atomizer* type simply sprays a fine mist of water into the air. In either version, the best models contain a humidistat, which acts like a thermostat to control the amount of moisture injected into the air.

Maintaining a Humidifier

Whether your humidifier is connected directly to the central heating system or is a portable room unit, it requires some basic maintenance.

Make certain the reservoir is full of water and that it is clean. Lime deposits in the water can coat the sides of the tank and should be cleaned off occasionally; dirt and dust can settle in the water and inhibit the unit's performance.

Most evaporator humidifiers have a pad, bristles, plate or brush that dips into the reservoir and carries the water past a blower fan. These items should be kept clean at all times, and replaced when they can no longer be thoroughly washed.

Inspect the water inlet valve for any deposits that might clog it and prevent water from flowing into the reservoir.

Check the float mechanism from time to time to be sure it is functioning properly and accurately reports when the water in the reservoir is low.

Keep the blower clean and its motor lubricated according to the instructions in the owner's manual.

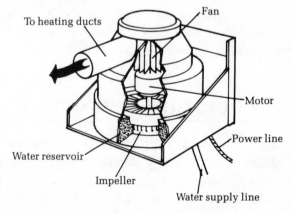

An evaporator-type humidifier that connects to a forced air heating system

6 COOLING systems

Until recently, air conditioning was considered a luxury. Room air conditioners, that is, portable units, were heavy and cumbersome, and needed special wiring circuits; central air-conditioning systems were only found in commercial buildings. But the state of home cooling has so evolved that now there are portable units that need no special wiring, and almost any home that is heated with a forced air system can use its duct system during the summer for central air conditioning of the entire building.

Nevertheless, air conditioning in any form is expensive. While the climate in many parts of the country makes the use of air conditioning almost mandatory, if you can live comfortably without it, you will save considerable money on your electric bill during the hot months of the year. One alternative to air conditioning is the use of fans. You can position fans in a window or two and in many instances that will provide sufficient comfort. If you have an attic fan, it will draw warm air out of the attic before its heat has a chance to filter down to your living space. If you have an attic fan mounted in the ceiling of the top floor, turn it on at twilight or when the weather is cool so that it can draw cool air into the attic at the same time it expels the hotter air under your roof.

With most forced air heating systems, there is a blower positioned at the return air side of the furnace which is used during the heating process to draw return air back to the furnace to be warmed. In hot weather, with the furnace turned off, you can run the blower by itself to circulate air in the house. This will not actually cool the air, but it will keep the air in the house moving, just as any fan will.

SELECTING AN AIR CONDITIONER

Figuring out how large or small an air conditioning unit you need requires some complicated computing, and the surest way of determining the proper size machine and where it should be positioned is to consult an air-conditioning engineer. Whatever it costs to get his advice will probably be saved in reduced electrical bills.

If an air conditioner is too large it will cool your home quickly and then shut off. Within a few minutes it will turn on again, consuming extra electricity in the process. Not only does this excessive cycling demand more electricity every time the unit starts up, but all that stopping and starting puts a heavy strain on its mechanisms, causing them to wear out sooner than they should. Conversely, a unit that is too small for the area it must cool will have to run continuously and even then it may not keep your home comfortably cool.

In order to determine the size unit you ought to have, a number of elements must be taken into account:

1•The square footage of the space to be cooled

2•The size and number of windows, the type of glass in each of them and the direction each of them faces

3•How the house is laid out in terms of its air flow and circulation

4•The R-value of all the insulation in the walls and attic

5•The type and position of the trees, shrubs and terrain that surround the building

6•The number and type of heat-producing appliances (including lights) that are used

When all of these factors have been considered, the decision comes down to the cooling capacity of the conditioner, which is expressed in British Thermal Units (BTUs) and its Energy Efficiency Ratio (EER), which defines how much the machine will cost you to run.

BTUs

All air conditioners have a BTU rating which defines the volume of heat that the unit can remove in a period of one hour. Some manufacturers calculate their units in tonnage, which means the number of tons of ice that must be melted to produce an equivalent amount of cooling. One ton equals 12,000 BTUs, so if you have a two-ton air conditioner it can remove 24,000 BTUs per hour. You can ignore the horsepower of an air conditioner. Horsepower is always listed on the unit's specification plate but machines with the same horsepower can produce significantly different BTUs, so it is a meaningless number so far as buying an air conditioner is concerned.

EER

After the size of a unit (the number of BTUs it can generate) has been arrived at, the next consideration is its energy efficiency. It is quite common to find air conditioners offering identical BTUs that have very different energy efficiency ratios (EER). Most places that sell air conditioners now have a conversion chart that allows the salesman to check the EER of any machine he has in stock. If there is no chart, the plate on any unit will give its wattage. Divide the wattage into the BTU rating to get the EER (BTUs ÷ WATTAGE = EER). The higher the number, the more efficient the machine and the less electricity it will consume.

For example, a 12,000-BTU machine might require 1,600 watts. Divide 12,000 BTUs by 1,600 watts and you get an EER of 7.5. But a similar model offering 12,000 BTUs might need only 1,200 watts and have an EER of 10. You would buy the second model even though it might cost a little more because the difference in its EER will pay for itself in energy savings.

CENTRAL AIR CONDITIONING

Both single-package and split-system units are used for central air-conditioning a home. The *single-package* type has its components housed in one big metal box, which is coupled to the furnace and uses the furnace's ductwork. It can be installed in the basement, next to the furnace, but the noise of the compressor is such that it is preferable to install it outside. If you put it outside, a connecting duct must be installed between the unit and the duct system inside the house, which means a big hole has to be cut in the side of your home.

The *split system* is by far the most popular central unit for home cooling. The split system involves installing the noisy compressor and the condenser coils on a concrete slab outside the house and placing the cooling evaporator coils inside, so that the noise

and vibrations of the machinery are removed from the living space altogether. The two components are connected by a pair of small-diameter copper tubes.

Installation of a central air-conditioning unit should be done by professionals. A competent installer will perform all the necessary operations to ensure the most efficient, least expensive system possible, including each of the items on the checklist below.

CENTRAL AIR-CONDITIONING INSTALLATION CHECKLIST

1 • The condenser must be located in a shaded area where it will receive no direct sunlight.

2 • The outside fan should not blow against any wall that will deflect discharged hot air back into the air intake. There must be at least 12″ between the air intake and any solid structure.

3 • The compressor is a large, heavy motor that not only is very noisy, but also vibrates a great deal. It should not be located near a patio or other exterior living area and should be properly mounted so that the vibration does not cause any internal problems. The vibrations, for example, are enough to loosen plumbing or electrical connections within the house.

4 • The duct system in the house should be large enough to handle the cool air traveling through them. Warm air travels more rapidly than cold, and if the heating ducts are too small the air condi-

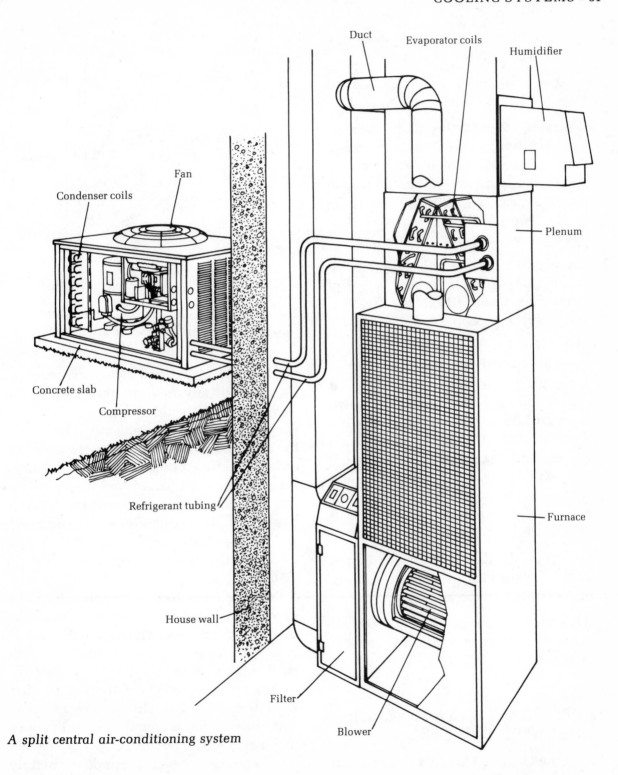

Duct

Evaporator coils

Humidifier

Plenum

Fan

Condenser coils

Furnace

Concrete slab

Compressor

Refrigerant tubing

House wall

Filter

Blower

A split central air-conditioning system

tioner will have to work harder and longer to cool your living space.

5 • All duct work should be insulated if it is to be used for air conditioning.

6 • Central air-conditioning units normally require 230 volts to run their motors and they must be given their own electrical circuit connected directly to the fuse or circuit-breaker box, so that other circuits in the house will not be overloaded when the air conditioner is running.

7 • The electrical cables and copper refrigerant line should be protected out of doors from damage from the weather, machinery, and people.

8 • Try to position the blower so that it can move air through the house, and also be accessible for cleaning. The blower and blower coil can be mounted in their own duct connected to the main heating duct. Or, if the blower is installed inside the furnace, the cooling coil must be positioned "downstream" so that cold and warm air will not meet inside the furnace and form condensation, which will corrode and rust the heating system.

9 • Be certain the thermostat is positioned so that it will give an accurate reading of the house's temperature.

10 • Be careful that the installation meets all the requirements set forth in your local building code.

MAINTAINING A CENTRAL AIR-CONDITIONING SYSTEM

Most of the working elements in an air con-

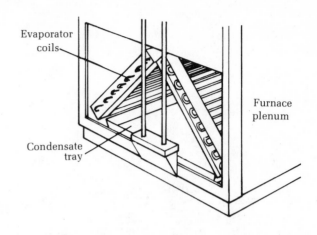

The evaporator coils are usually attached to the furnace plenum and should be kept clean

ditioner are sealed and cannot be reached for repair by anyone other than a qualified serviceman with the proper knowledge, gauges, and other equipment to deal with the enclosed refrigerant, freon. However, in order to perform at its maximum efficiency, every air conditioner must be kept clean, which means its filters should be changed frequently. Where the filters are positioned and the procedure for cleaning or changing them are described in the unit's owner's manual.

Both the condenser and evaporator coils must also be kept free of dirt and should be cleaned periodically. The evaporator coil is located in the plenum, at the top of the furnace. If the plenum is sealed the evaporator cannot be reached. But if the plenum has insulation around it, trip the circuit breaker that controls the air conditioner and remove the insulation. The insulation is usually taped to the metal plenum; undo it carefully so that it can be reused. Undo the retaining screws that hold the access plate over the evaporator coil and reach inside with a suede brush to clean the underside of the coils. You may be able to pull the unit forward a few inches, but be careful not to bend the pipes connected to it. Also clean the tray located beneath the evaporator unit. The tray is supposed to carry away any condensation that collects on the coils and it will become clogged if it gets dirty. You will find a small hole in the tray; pour household bleach into the hole to get rid of any fungus that may be developing in the drain.

To clean the condenser coils in the outside unit open the circuit breaker to the unit. Cut down any weeds or grass that have grown up around the condenser and might obstruct the air flow, and then remove the grille that covers the coils (some units do not have a grille). The fins on the condenser are made of aluminum and they should be cleaned with a brush. Be very careful not to bend or break any of them.

ROOM AIR CONDITIONERS

Room air conditioners are often called portable units. And so they are—if you have two large men and a small orangutan to move them. They come in many sizes and designs and are built to cool anything from one room to a small house. Most of the smaller units can now be plugged into a standard 120-volt wall outlet; the larger, more powerful models require their own 240-volt circuit.

When you are buying a room air conditioner you want one that is neither too

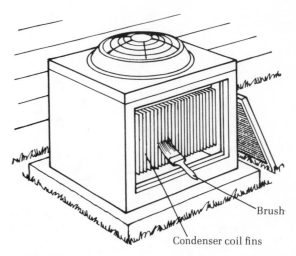

Brush

Condenser coil fins

Clean the condenser coils with a brush. Do not bend the coil fins

small, nor too large. The number of windows and the direction in which they face are important details to consider; the number of doors in the room, the room's insulation, which floor the room is on, and the number of heat-producing appliances in the room must all be taken into account. Once you have decided what size (number of BTUs) you need, look for the unit that will give you the highest EER. It should be noted that if you are trying to cool more than one room with a room air conditioner you can assist in the circulation of cool air by situating a fan or two opposite the air conditioner to help draw cool air through the house.

INSTALLING A ROOM AIR CONDITIONER

Room units are meant to be installed by their owners, although you can hire a professional to do the job. Most units come with installation instructions, but the manufacturer usually sells a mounting kit as

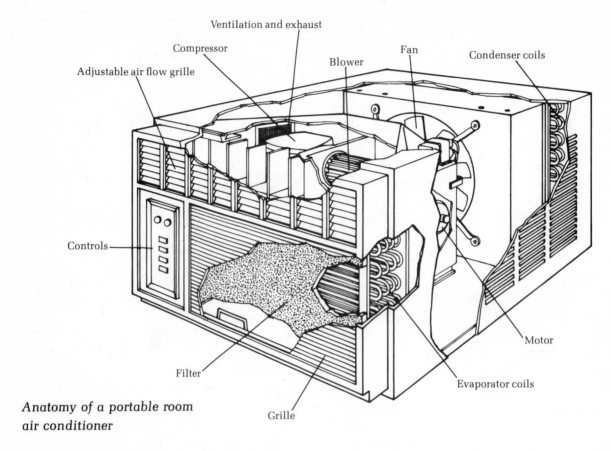

Anatomy of a portable room air conditioner

well. The kit includes a metal frame called a cradle, side panels to fill in the spaces between the unit and the window frame, and gaskets for sealing the unit. You have the option of installing a room unit through al-

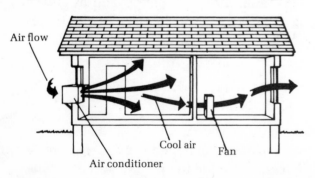

Cool air will flow through a house better if it is helped by an ordinary room fan

most any wall, in which case the side panels are not necessary.

Window Installation

To install a room air conditioner in a double-hung window, the steps are as follows:

1 • Select a window to hold the air conditioner. Make sure you do not need to open it during the cooling season, and that it does not receive any direct rays from the sun during the better part of the day. If you have no choice in the window you use, consider putting an awning over it to shield the unit. The back half of the air conditioner containing the condenser

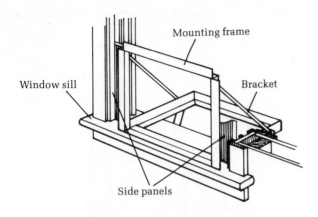

Assemble the air conditioner mounting bracket

will project outside, where it must have good air circulation so that the heat drawn from inside the house can dissipate; the cooler the air around the condenser, the more efficient the unit will operate.

2 • Assemble the mounting frame and attach the cradle to the windowsill, following the instructions that come with the mounting kit. The unit must slant approximately ¼″ toward the outside of the house so that the condensate that forms

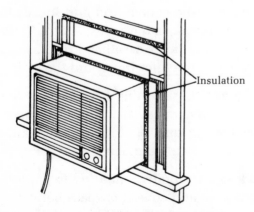

All air gaps around the air conditioner must be filled

inside the conditioner can drain out of the machine.

3 • When the cradle is mounted firmly, attach the side wings to the frame and apply weather stripping to the bottom of the lower window sash and to the inside of the lower sash's top rail, to fill the gap between the lower sash and the glass in the upper sash.

4 • Place the unit in its cradle (you'll need a friend with muscles to help) and secure it according to its instructions.

5 • Install gaskets around the unit; seal every air gap found around the unit.

6 • If the wiring for your air conditioner must be a special branch circuit, be sure to install it according to your local electrical code.

Wall Installation

You can also install an air conditioner in an outside wall. If the exterior of the wall is metal siding or brick, you are in for some extra hard work, so you might want to reconsider the project, but it is feasible to accomplish. Air conditioners installed in walls have a special housing, which is inserted through the wall to hold the unit. The general procedure is this:

1 • Measure the size of the housing and mark it on the inside wall, preferably between two studs (there will be another stud between the two).

2 • Drill into the wall at one of the corners of your outline and saw out the hole. You will encounter at least one stud, but don't cut if off yet. You will presumably

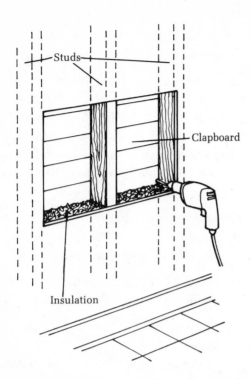

To install a wall air conditioner cut out the inside of the wall first. Use 1″ × 4″ boards to keep loose fill insulation from draining out of the hole

Frame the hole with 2″ × 4″ or 1″ × 4″ boards

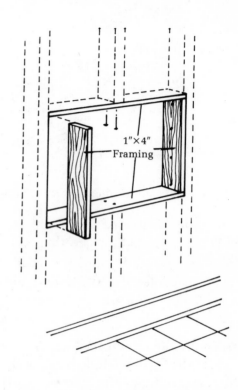

also encounter a lot of insulation. If the insulation is batts or blankets, cut through it. If it is loose fill, cut a 1″-wide slit along the top of your outline and insert a 1″×4″ board into the slit on both sides of the center stud to hold the loose fill in place. Then cut away enough of the wall below each end of the board to drive nails into the board and hold it in place. You will have the same problem along the sides of the hole and need to insert similar boards to stop the fill from leaking out of the wall.

3 • When you have cut through the inside wall to the outside sheathing, drill through each corner of the hole. Then go outside and remove whatever shingles or clapboard are between the holes and saw through the sheathing.

4 • Remove the center stud and install filler studs between the walls, using 2×4s or 1×4s. Caulk or weather-strip the space between the filler frame and the walls.

5 • Remove the housing from the air conditioner and install it in the wall opening so that it slants ¼″ downward toward the outside. Caulk the housing.

6 • Make a decorative molding to go around the hole and cover the space between the housing and the inside wall; attach it around the housing.

7 • Insert the air conditioner into its housing and secure it.

8 • Check the installation for air leaks and caulk any you find.

MAINTAINING A ROOM AIR CONDITIONER

Most of the essential elements of any air conditioner are sealed to contain the refrigerant, freon, which moves from the evaporator coils to the condenser coils and back again. Any problems that develop with these components should be repaired by a qualified serviceman. But there is some routine maintenance you can do to keep your air conditioners operating at peak efficiency:

1 • Vacuum the condenser and evaporator coils occasionally to get rid of any accumulated dirt or dust.

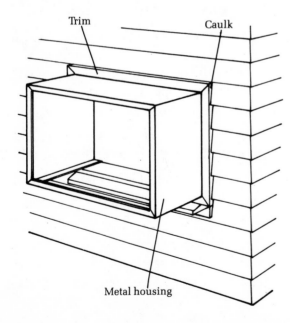

Trim

Caulk

Metal housing

Insert the air conditioner housing in the wall

2 • Clean the drain tubes that carry the condensate out of the machine by poking a straightened coat hanger into the tubes.

3 • Inspect the evaporator and condenser coil fins. If any of them are bent, straighten them out. You can use a fin comb, pliers or your hands, but remember they are delicate, so be careful not to break them.

4 • Clean or replace the air filter at the beginning of each cooling season.

THINGS TO DO BEFORE CALLING A REPAIRMAN

When your air conditioner fails to operate properly, there are several preliminary checks you can make to be sure the failure is internal and not something obvious, such as a loose line cord.

If you turn on the unit and it does not start: Check the fuse or circuit breaker that controls the unit's circuit. If the fuse has not blown or the circuit breaker is still on, check the wall outlet and the unit's plug. The power cord can be tested with a multimeter.

If the unit blows warm air: Either the thermostat or the compressor is defective. Check the thermostat with a multimeter. If it has continuity, the compressor is bad and you will have to call a repairman.

If the unit does not blow enough cool air: Check the fan. Its blades may be bent or the fan motor may need lubrication. Be certain the setscrew that holds the blower to the back of the motor shaft is tight. If all of this fails, change the filter, clean the evaporator and condenser coils, and straighten any bent fins you encounter.

If the compressor operates erratically: Examine the sensor bulb to be sure it is positioned in the return air flow and not touching any of the coils. Be sure no drapes or other obstructions are blocking the condenser coils. Clean the condenser coils with a vacuum.

If the unit is very noisy: Check the fan blades and motor bearings. The blades may need cleaning or straightening and the motor may need to be lubricated. Be sure the setscrew holding the blower is tight. The machine may just be loose in its mounting. Tighten the mounting bolts.

If water drips inside the house: Reposition the unit so that it tilts outward; clean the drain tubes. If the weather is really humid, install a drain pan under the unit.

RUNNING AN AIR CONDITIONER ECONOMICALLY

There are several things you can do to conserve the energy your air-conditioning system uses and keep the cost of its operation to a minimum:

1•Turn off lights not in use. Light bulbs give off tremendous heat, forcing the air conditioner to work harder.

2•Keep the room or house temperature no more than 20° below the outside temperature.

3•Keep heat-producing appliances away from the thermostat so that it can give accurate readings.

4•With a central cooling system, run the blower fan continuously to reduce the amount of time the cooling system has to be in operation.

5•Clean or replace all filters at least once a year.

6•If you have attic or window fans, do not run them at the same time as your air conditioner.

7•Use your kitchen vent fan whenever you are cooking to keep heat from building up in the house.

8•High humidity makes any air conditioner work harder and longer. So cook, bathe and launder before or after the air conditioner is in operation.

9•If you have a bathroom exhaust fan, use it after bathing.

10•Be sure all room air conditioners are shaded; outside compressors should also be protected from the sun.

11•Keep the air conditioner clean and not blocked by drapes or furniture.

12•If you have forced air heating but are using window air conditioners, close or block off the air return grilles or you will wind up cooling the basement instead of your living areas.

13•Shut off your air conditioner whenever you leave home for more an hour or two.

14•Use shades or blinds over all windows to keep the sun's direct rays from entering your home.

15•When the air conditioner is operating, keep the windows closed and try not to open doors any more than is necessary.

7

HOT WATER,
APPLIANCES
and LIGHTING

Within your house there are a surprising number of modern appliances, all designed to save you time, space or effort. They are marvelous conveniences that wash and dry the laundry and dishes better than you can do them by hand. They can preserve or cook the food and heat the water, give you light, and even entertain you when you want to be amused. But their performance comes at the high cost of energy, which is usually in the form of electricity.

You cannot avoid the expense of running modern appliances unless you stop using them altogether. But you can take steps to insure that they function efficiently, with a minimum of wasted energy. Obviously, the appliances themselves must be maintained in good working order and be repaired as soon as anything goes wrong with them. But there are still things you can do to help even the newest appliance use only the energy it needs to satisfy your living comfort.

HOT WATER HEATERS

The largest single energy consumer in any home is the heating system. The second largest energy consumer is the hot water heater. The hot water heater is that large can tucked away in a dark corner of your basement which rarely causes any trouble. It may well go on heating your water day in and day out for 20 years before anything happens to remind you it is even there. The task of the water heater is to heat 50 or more gallons of cold water and keep it hot for any moment you open a hot water tap somewhere in the house. Actual heating of the water may consume natural gas, oil or elec-

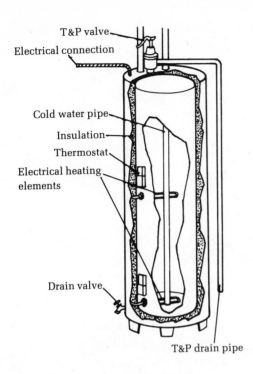

An electric hot water heater

valves to be sure they are operating correctly. In particular, lift the lever on the temperature/pressure (T&P) valve in the top of the unit. If the valve hisses, it is functioning. If there is no steam, water or noise, replace the valve. The purpose of the T&P valve is to release excessive pressure from the tank whenever the temperature of the water goes beyond its predetermined limits. Without a T&P valve, should there be a failure in the heating system, conceivably the tank could explode.

2•Drain the heater once a month to remove any sludge or deposits that accumulate at the bottom of the tank. There is a faucet at the base of the heater. Open it and drain the water into a pail until it becomes clear. If you drain regularly once a month you will probably have to take off no more than a gallon or so of water each

tricity. Maintaining the water's temperature is a function of the heater's insulation, which is packed between its outer shell and a watertight container. Touch the outside of your water heater. If it feels warm the water in it is losing valuable heat. Because the heater is controlled by a preset thermostat, it knows that it is supposed to keep its contents at about 140°F at all times, day and night. When the water drops below the given temperature the unit will heat it up again.

MAINTAINING A WATER HEATER

While a water heater has no moving parts, it still must be cared for with some regularity.

1•Periodically check the thermostats and

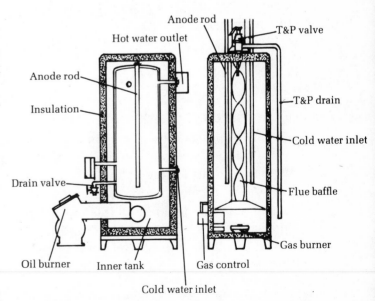

An oil-fired (left) and a gas-fired hot water heater

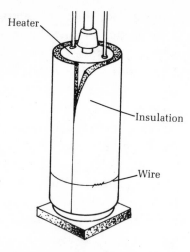

Insulation can be held around a hot water heater with tape or wire

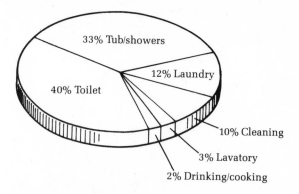

How a typical family spends the 250 gallons of water it uses each day

time. If you do not drain the tank at all, eventually mineral deposits from the water will eat through the bottom of the tank and you will have to buy a new heater.

3 • If the heater is gas or oil-fired, check and clean all of its burners and flues at least twice a year.

SAVING ENERGY WITH HOT WATER HEATERS

1 • If the outside of your hot water heater is warm, wrap the entire unit with a blanket of insulation; you can buy special heater insulation kits at most home building centers. Be careful not to cover any vents in the top of the unit. If the tape that comes in the kit loses its bonding qualities after a few months, hold the insulation in place with lengths of wire.

2 • The temperature setting in your hot water heater should be set at medium, or approximately 140° to 160°F.

3 • If you go away for more than a few days, turn off the heater.

4 • Consider installing a timer on your heater which can be set to shut down the unit at night and then turn it on again an hour or so before you normally get up in the morning. Unless you are in the habit of taking hot showers in the middle of the night, there is no reason to spend fuel or electricity to keep the water hot for the eight hours out of every day when no one will use it.

5 • You might also be able to reduce your water heating bill by installing a small water heater in areas where you need 150° or 160°F water, such as in a laundry room, or near your dishwasher. Such an installation can save on the cost of heating your water by allowing you to maintain a 120°–140° temperature in your main heater because it will only be serving such general uses as bathing and cooking.

SAVING WATER

You pay for all the water you use. If there is a meter attached to the water main that en-

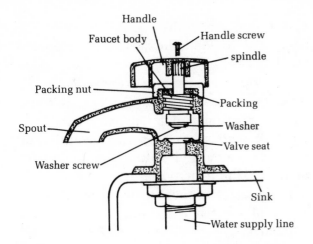

To replace a faucet washer in a dripping faucet: 1) Undo the handle screw and pry off the handle. 2) With a wrench, unscrew the packing nut. 3) Unscrew the faucet body, or pull it out of its housing. 4) Undo the screw holding the faucet washer in place and replace the washer. 5) Reassemble the faucet by reversing the order and following steps 4 thru 1. If the packing under the packing nut is compressed, replace it. Rather than packing there may be a washer. If that is worn, replace it.

ters your house, you pay for every drop that comes into the building. If there is no meter in your house, you still pay an annual municipal tax. So it makes sense to get the most from the water you receive and to waste as little of it as possible.

An average American family of four consumes 250 gallons of water every day of the year. Add to that even the smallest unnecessary waste, and water becomes very expensive indeed. For example, a single dripping faucet can drain off about $50 worth of water a year. That is a ludicrous expense when you consider that a leaky faucet can be fixed by anyone with a pipe wrench and a two-cent piece of rubber in less than half

an hour. A plumber will fix it for you at a cost of about $25 and you could still save money. You might also replace all of your faucets, including the shower head, with low-flow attachments. Any of the low-flow attachments on the market today will curb water consumption and still provide ample water for bathing or cooking.

Then there is the toilet, which accounts for 40% of the water consumed each day. A great deal of water (and money) can be saved by simply not flushing the toilet each time you dump nothing more than a cigarette butt or toilet tissue in the bowl. Here again, you can install any of the many new inexpensive water-saving flush devices available at hardware and home improvement stores. And if you are remodeling or adding a bathroom, consider one of the new water-saving toilets. Do not put a brick in the tank to raise the water level. It will not save much water and it may disintegrate and damage the toilet or clog your drain-vent-waste line.

Look at the hot water pipes that emanate from your water heater and go across the basement ceiling. They should be wrapped in insulation. Hot water loses one degree of heat for every foot of uninsulated pipe it travels. That is not particularly critical if the water is coming out of a bathroom or kitchen tap where it will be used for bathing or cooking. But the hot water feeding your dishwasher or washing machine must arrive at the appliance at a prescribed temperature. For example, hot water entering a clothes washer should be at least 140°F. If the hot water heater is set at 140°F and the washer is 50 feet of uninsulated copper pipe away, the water will only be 90°F when it reaches

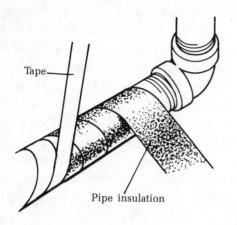

Wrap all exposed hot water pipes with insulation

the machine. To have the proper water temperature at the washer, you either have to insulate the pipes or set the hot water heater up to 190°F. Not only will a 190° setting waste tremendous amounts of energy, but when it comes out of the tap in your bathroom at a little more than 22° below boiling, somebody is bound to be scalded.

Wrap exposed hot water pipes wherever you can get at them, particularly under crawl spaces where they are exposed to cold air. There are a variety of easy-to-install insulating products on the market for encasing water pipes.

APPLIANCES

When you are finished eyeing your hot water heater as a consumer of energy, turn your attention to the appliances in your home. The small ones use up considerable amounts of electricity every time you use them, and there is very little you can do to reduce their voracious needs. But the amounts of electricity your toaster, coffee maker and steam iron need are nothing compared to the demands of your major appliances—the refrigerator, dishwasher, oven, clothes washer and dryer. Because the major appliances are so large, there are some small repairs and maintenance chores that you can perform to keep them running at peak efficiency and reduce energy waste. Some of these activities are mechanical; most are a matter of habit, or at least of being aware of what it costs you to operate the appliance.

REFRIGERATOR-FREEZERS

Misuse your refrigerator and it will cost you a fortune on top of the huge amount of electricity it normally consumes. The purpose of a refrigerator-freezer is to preserve food

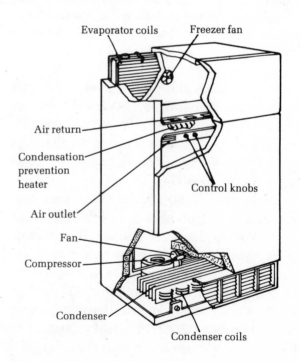

A refrigerator-freezer

in a state of coldness. The appliance functions much as an air conditioner does, except instead of blowing the cold air it manufactures into the room, it contains it inside. A refrigerator-freezer is a rather jealous beast about all the cold it creates, so the moment its temperature goes below 34° to 37°F, its motor starts up and it begins making more cold. Every time you open a refrigerator or freezer door, cold air escapes. If there is a leak between the door and its frame, more air escapes. Cold air is forever escaping from a refrigerator-freezer. And the machine is forever crying out for more electricity to run its motor so it can replace its losses.

Like all appliances, a refrigerator-freezer operates most efficiently when it is clean. Not the front, so much as the insides and the condenser coils. Vacuum the coils at least every three months, using a brush attachment. Considerable dust will collect on the coils and reduce their cooling efficiency, which results in the appliance running more often and for longer periods of time than it needs to. Ice must also be cleaned off the freezer compartment if the unit is not a self-defrosting one. Keep an eye on the frost buildup in the freezer compartment and defrost it whenever it exceeds a ¼″ in depth, or it will begin to act like an insulator and reduce the freezer's efficiency.

The temperature in the freezer compartment should read between 0° and 5°F. Place a thermometer in the freezer. If it reads below 0° or above 5°, reset the freezer thermostat. The temperature in the refrigerator compartment should range between 34° and 37°F. If a thermometer placed in the food compartment does not read within that range, reset the compartment's control.

Perhaps the most common malady that occurs with refrigerator-freezers is air leakage around the door gasket. Hold a dollar bill against the door frame and close the door on it. If you can pull the bill out of the door with ease, the gasket is not sealing properly, particularly if you try the bill at different positions around the door and meet little or no resistance when pulling.

The door hinges may have loosened, so check them for loose screws. More often, the gasket has become worn or damaged and should be replaced. You will have to purchase a manufacturer's replacement and at the same time get some advice on exactly how the gasket on your model is attached to the door. As a rule, gaskets are held in place by screws that go through a retainer strip, the gasket itself, the inner door (which has the shelves molded in it) and finally the outer door. Although installation of the gasket varies from model to model, you do not want to remove all the screws at one time or the inner door will come off in your hands. Peel back the old gasket to remove the screws along the top and then screw the new gasket in place before you do the sides and bottom. When the gasket is in place, smooth it out with your hands and test it in several places with a dollar bill.

Saving Refrigerator-Freezer Energy

1 • Open refrigerator and freezer doors as seldom as possible, particularly during hot weather.

2 • Maintain the proper temperature in both the refrigerator and freezer compartments. It is not necessary to have the

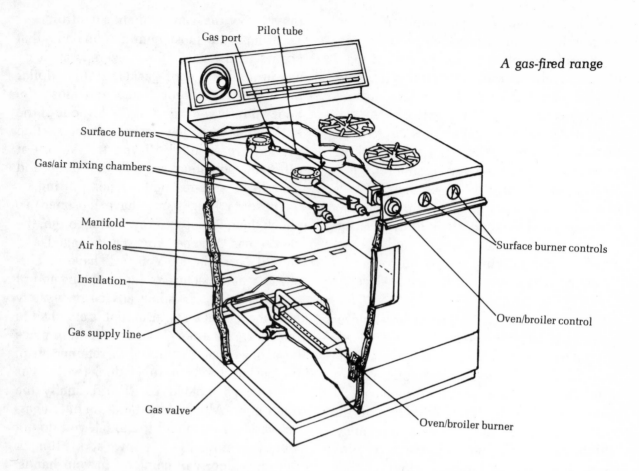

A gas-fired range

Gas port · Pilot tube

Surface burners

Gas/air mixing chambers

Manifold

Air holes

Insulation

Gas supply line

Gas valve

Surface burner controls

Oven/broiler control

Oven/broiler burner

freezer any colder than 0° to 5°F or the refrigerator more or less than 34° to 37°F.

3 • Defrost the freezer regularly.

4 • Vacuum the condenser coils every three months.

5 • If you have a frost-free unit, tightly cover any liquids you put in it. Liquids tend to evaporate, which makes the frost-free system work more than it should.

6 • When you go on vacation, either unplug the refrigerator-freezer or raise its temperature setting so that it will not turn on as often, but can still keep whatever food is left in it from spoiling.

7 • Try to keep the freezer full so that the frozen food can help to retain the cold. Conversely, avoid loading large amounts of unfrozen food in the freezer all at once. The machine will have to work doubly hard to stay cold.

8 • Do not pack the refrigerator compartment so tightly that air cannot circulate.

9 • Position your refrigerator-freezer away from such heat-producing appliances as the oven and make sure there is plenty of space behind the unit for air to circulate.

OVENS, STOVES AND COOK TOPS

Whether it is gas-fired or electrically powered, a major cooking appliance uses up almost as much energy cooking food as the refrigerator needs to preserve it. The average family of four spends about 240 hours a year cooking. It goes without saying that if your cooking unit is not kept clean and in good working order it will waste a considerable amount of energy during those hours. For cleaning and maintenance follow the instructions found in your owner's manual.

Meanwhile, here are some ways you can use your oven and surface burners less and conserve the energy you spend on preparing food:

Saving Cooking Energy

1 • Cook with as little water as possible. The less water you have in a pot the quicker it will boil.

2 • Liquids boil quickest in a tightly covered pan and will save about 20% of the energy you would use otherwise. As soon as the liquid comes to a boil, lower

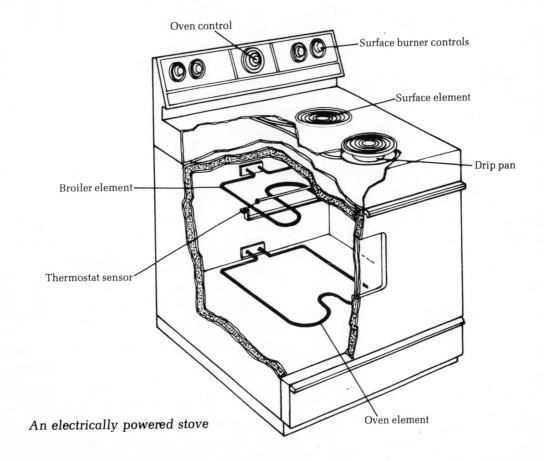

An electrically powered stove

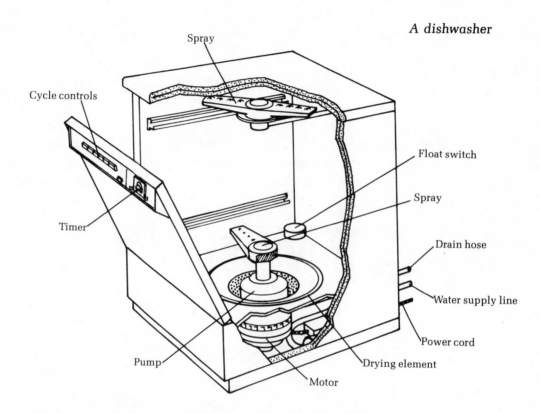

A dishwasher

Spray

Cycle controls

Float switch

Spray

Timer

Drain hose

Water supply line

Power cord

Pump

Drying element

Motor

the heat enough to keep it just bubbling. Any higher setting is a waste of energy.

3 · Thawed foods cook faster than frozen ones.

4 · Keep the bottoms of your pans shiny. A dirty pan or pot lessens its heating efficiency.

5 · Copper, stainless steel and cast-iron pans need less heat than aluminum. If you use a glass or ceramic pan when baking, set your oven 25° lower than with a metal container.

6 · Pots or pans should be the same sizes as the burners you put them on; a small pot on a big burner only wastes electricity.

7 · Don't turn on a surface element until the pot is on top of it. The only thing you will cook is thin air—at a very high price for energy.

8 · Use your toaster, toaster-broiler, crock-pot, electric frying pan, any of the small cooking appliances whenever you can. They use less energy than your stove's surface burners. When you have to use the stove, remember that a pressure cooker will reduce cooking time and save money.

9 • An electric burner will take as much as five minutes to cool down after it is turned off, and a gas burner stays hot for at least a minute. So shut off the burner just before the food is done and let it finish cooking as the burner is cooling.

10 • An electric burner uses 1.5 kilowatts an hour. An oven consumes over 5 kilowatts. Use the burners for fast-cooking foods and the oven for foods that require longer cooking.

11 • Keep the reflectors under the surface burners clean and shiny so they can efficiently reflect heat up to your pots.

12 • If you have a microwave oven, use it. It demands a lot of energy, but for very short periods of time.

13 • Never use an oven or stove for heating the kitchen.

14 • Try not to preheat the oven, and never preheat the broiler. If you must preheat the oven, ten minutes of wasting energy and dollars is enough.

15 • Anytime you use either the broiler or the oven, try to cook as many dishes as you can at the same time. Dishes with cooking temperatures of plus or minus 25° can be cooked together.

16 • If you cook an oven roast at 325°F the meat will shrink less, splatter less grease, and use less energy.

17 • Don't keep opening the oven door while you are baking. You lose 20% of the cavity's heat every time the door is opened.

18 • If you have double ovens, use the small one whenever you can. It demands less energy.

19 • Never line the inside of an oven with aluminum foil. It will reduce the oven's efficiency and may cause the heating elements to burn out.

20 • If you have a self-cleaning oven, put it into the cleaning mode immediately after cooking, when the oven is already a long way toward the 900° temperature it must attain for cleaning.

21 • Use any leftover heat in your oven to warm rolls or dishes.

DISHWASHERS

A dishwasher actually uses less water than you would to clean the same number of dishes by hand. It also spends a considerable, but not excessive, amount of electricity. As with any major appliance, it pays to keep your dishwasher in good working order; dishwashers are so energy efficient there are only a few things you can do to save on the cost of running them.

Saving Dishwasher Energy

1 • Run the dishwasher only when it has a full load.

2 • You do not have to prerinse your dishes if the machine is a relatively new model. However, if the dinner plates are going to wait a day or so for the dishwasher to be filled, rinse off starchy foods and bits that may harden.

3 • Load the machine properly, that is, according to the manufacturer's suggestions in the user's manual. If you load your dishwasher in any other way you will inhibit its cleaning efficiency.

4•Shut off the unit as soon as it reaches its drying cycle, and open the door to let the dishes air dry. A lot of electricity gets used in drying dishes when they could dry all by themselves for free.

5•Keep the filter screen free of food particles. It should be cleaned once a week or so.

6•Use the proper amount of a dishwasher detergent. Too much or not enough detergent will reduce the unit's efficiency.

CLOTHES WASHERS AND DRYERS

A clothes washer uses large amounts of water and electricity; dryers all need electricity to some extent, and some of them also rely on natural gas to generate heat. Recent versions of both appliances are designed to function on minimum energy demands, providing they are properly maintained. You can save on your hot water bill if the pipe running from your hot water heater to the washer is as short as possible, or at least well insulated. Similarly, the dryer should be vented outdoors with no more than a 10′ run of tubing.

Saving Clothes Washer and Dryer Energy

1•Delicate fabrics require shorter washing and drying cycles than heavy work clothes; sort your washing and drying loads according to the fabrics.

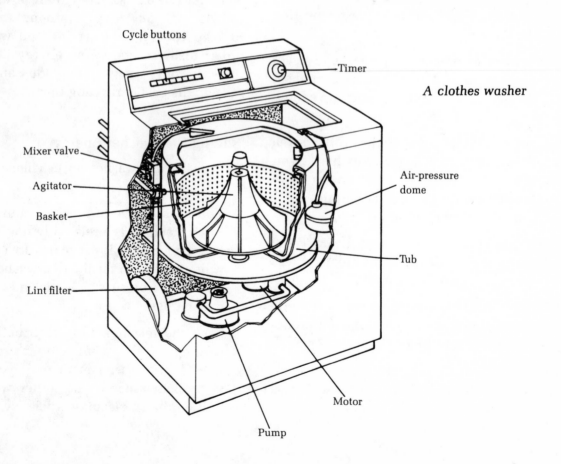

A clothes washer

Cycle buttons

Timer

Mixer valve

Agitator

Basket

Lint filter

Air-pressure dome

Tub

Motor

Pump

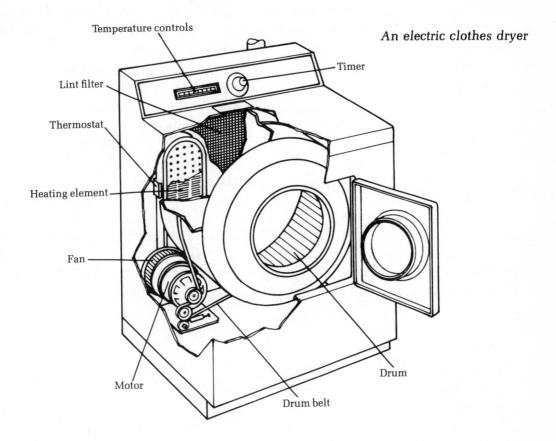

An electric clothes dryer

Temperature controls

Timer

Lint filter

Thermostat

Heating element

Fan

Motor

Drum belt

Drum

2•Never use more water in a wash than you have to. On the other hand, don't overload your selected water level or the washer will have to work harder. If your washer has no water level selector, wait until you have a full load before washing. It costs you about 5 kilowatt hours of electricity every time your washer goes through a full cycle. As a rule of thumb, wait until you have a full load before using your dryer. But don't overload it, either. The clothes will take longer to dry and the machine will have to work harder.

3•Clean the lint filters in both the washer and dryer after each time they have been used.

4•Dry clothes in consecutive loads. The dryer retains much of its heat and it will have to work less going directly from one load to the next.

5•Use only the proper amount of detergent for any wash. Oversudsing will cause the washer to overwork and doesn't get your clothes any cleaner.

6•Set the timer on your dryer to the minimum length of time needed to dry its load. Overdrying can wrinkle some fabrics and costs money in energy usage. If you plan to iron the clothes you are drying, and you will have to dampen them anyway, don't dry them completely to begin with.

7 • If you have a laundry tub which you use when draining your washer, consider using the sudsy water for cleaning garden tools, mops, etc.

LIGHTING

Roughly 22% of your annual electrical bill pays for the lights you turn on every night, and in most cases you can reduce that electrical consumption without giving up either light or comfort. If you are interested in reducing your lighting bill, the first thing you have to do is stop thinking about light bulbs in terms of their wattage, and start considering their lumens.

The wattage marked on most incandescent bulbs is a measure of how much electricity the bulb uses. For 80 years now Americans have assumed that the more light they need, the higher the wattage of the bulb should be. In essence, that is true—sometimes. The most important fact about a light bulb is how much light it produces, that is, its brightness, which is measured in lumens. The lumens are sometimes printed on the bulb itself, but more often you will find them stated on the package, in a nice, big number like 2,880 (for a 150-watt bulb) or

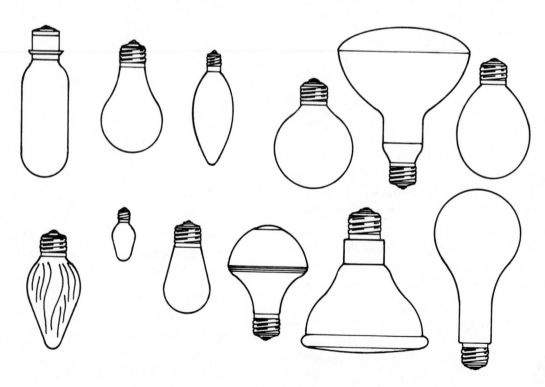

Incandescent bulbs come in many shapes and sizes

1,170 (for a 75-watt bulb). Note that two 75-watt bulbs offer a total of 2,340 lumens, which is 540 lumens less than you get from a 150-watt bulb. As a matter of fact the difference in lumens is such that you would need to turn on six 25-watt bulbs to get the same amount of lumens that a single 100-watt bulb will give you—but you would be paying for 150 watts of electricity to achieve the same number of lumens you can have for the price of 100 watts. So the first rule to remember about lumens is that incandescent bulbs of lesser wattage are less efficient than larger wattage bulbs.

Incandescent bulbs sold as *long-life bulbs* are able to last longer than regular bulbs because they put out less light. In other words, a long-life bulb has a smaller lumen measurement than a standard bulb of the same wattage. Since the two bulbs consume an identical amount of energy (if their wattages are the same), a long-life bulb is not used to save energy. But if it is difficult or inconvenient to change bulbs, or if the amount of lumens needed is less than you would get from a standard bulb, then a long-life bulb has the advantage.

SAVE ENERGY WITH INCANDESCENTS

Incandescent bulbs come in all manner of shapes, sizes, colors, wattages and lumens, so you could go around your house and put in bulbs that will deliver only the amount of lumens you need for the various activities that go on around each light. If that is your persuasion, a general breakdown of the number of lumens required for different activities follows:

TASK	LUMENS
Reading, writing, studying	70
Playing cards, billiards, table tennis	30
Shaving, combing hair, applying makeup	50
Working in the kitchen	50–70
Laundering, ironing	50
Sewing (detailed), low-contrast dark fabrics	200
Sewing, light- to medium-colored fabrics	100
Sewing for short periods of time	50
Close handicraft work (reading blueprints, diagrams, fine finishing)	100
Cabinetmaking (planing, cutting, sanding, gluing, measuring, assembly, repairs)	50
Any area involving a visual task	30
Any passage area or area used for relaxation or conversation	10

A more reasonable, and certainly more versatile, way of reducing the lumens you use to the minimum you need for different occupations is to install either three-way bulbs or dimmer switches. Three-way bulbs offer three different lumens, each at a different wattage, such as 50-100-150, and the glory of them is that you can start by turning the bulb on at its lowest wattage, then go to a higher intensity only if you need it. Three-way bulbs do need a special socket, which has three switching positions, but these are the same size as any standard socket and are easily installed:

Changing a Light Socket

1 • Unplug the lamp and remove the bulb from its socket.

2 • The socket shell is a metal tube inserted into a base. Pull the tube upward to disengage it from its base and remove the cardboard sheath that protects the wire connections.

3 • Undo the two terminal screws on the side of the socket and remove the socket wires around them. There may be a setscrew in the socket base that must be loosened before the base can be unscrewed and pulled off the wires.

4 • Disassemble the three-way socket and thread the wires in the lamp through the socket base.

5 • Screw the base to the lamp and tighten the setscrew.

6 • Wrap one wire clockwise around each loosened terminal screw, then tighten the screws.

7 • Insert the cardboard sheath and socket housing over the wires and into the base. Twist the shell slightly until it locks.

8 • Put a bulb in the socket and plug in the lamp to see if it works. If the bulb does not light, you did not make the proper wire connections. Disassemble the socket and rewrap the wires around their terminals; tighten the terminal screws as much as you can.

Dimmer Switches

The alternative to using three-way bulbs is to install a dimmer switch. Dimmers can be purchased for attachment to a lamp cord or to replace a standard toggle switch situated in a wall, and they cost between $5 and $15. The great advantage of dimmers is that you can control exactly how much brightness you get from any light bulb, and the less brightness (fewer lumens) you use, the less money you spend per kilowatt hour. The initial cost of a dimmer attached to a light you use daily will be repaid in savings on your electric bill within a year's time. As for installing a dimmer, it is almost easier to do than changing a socket.

Installing a Dimmer

1 • Turn off the circuit breaker or remove the

The components of a light socket

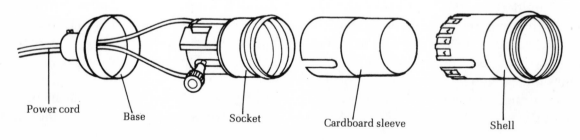

Power cord   Base   Socket   Cardboard sleeve   Shell

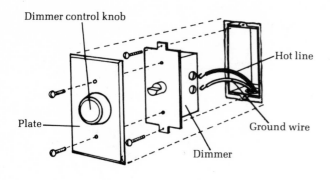

How to wire a dimmer

fuse controlling the light switch you intend to change.

2 • Remove the faceplate over the switch.

3 • Undo the top and bottom screws that hold the switch in its electrical box. Pull the switch out of its box. The cable wires attached to the switch are stiff, so you may have to pull pretty hard.

4 • Undo the terminal screws on the sides of the switch and release the cable wires. You can now remove the switch.

5 • Dimmers are attached to the cable wires in various ways. Follow the installation instructions that come with the dimmer precisely. They will tell you exactly where and how to attach each wire.

6 • When the dimmer is connected, push it into the electrical box and hold it in place by tightening the top and bottom screws in the electrical box. Attach the faceplate and knob and then turn on the electricity.

FLUORESCENT LIGHTING

You probably already have some fluorescent lights in your kitchen and/or bathroom and you might consider adding more of them as a step toward reducing your light bill. Fluorescents offer the cheapest source of lumens you can buy. They use 75% less energy than incandescent bulbs of the same wattage; a 40-watt fluorescent will give you more light than a 100-watt incandescent, at less than half the cost in electricity.

There are three kinds of fluorescents manufactured today: preheat, rapid start, and instant start. All three types can be purchased either as tubes, ranging from 1' to 8' in length, or as rings 6" to 2' in diameter. They all also come in a range of colors, each of which delivers a different quality of light.

The big complaint with fluorescent lamps is that they seem to flicker, and until very recently the shade of light made many people look weird, as well as put a strain on their eyes. Modern fluorescent bulbs such as "soft white" have corrected those complaints. Another problem with fluorescent lights is that they require special fixtures, which are relatively expensive, so it would cost considerable money to change all incandescent lamps in your house over to fluorescents, even though the fluorescents would pay for themselves in reduced electric bills. Consider fluorescents whenever you are planning to install additional lighting, however, and most certainly use it if you are putting in a luminous ceiling or want to light a large work area, such as a workshop or kitchen. Fluorescent bulbs, by the way, will last as much as 20 times longer than incandescents.

HIGH-INTENSITY DISCHARGE LIGHTING

You cannot use the high-intensity discharge

(HID) bulbs and fixtures in every lighting situation. But bear in mind that they provide from two to five times as much light as an incandescent bulb of the same wattage, and will last from 10 to 30 times longer. The high-intensity lamps use mercury and metal halide bulbs, which can direct a high intensity beam of light to small areas. They are superlative for reading or lighting small work areas without running up your electric bill.

SAVING ENERGY WITH LIGHTS

1 • Turn off incandescent lights whenever they are not needed, even if only for a few minutes. Fluorescents should be left on whenever turning them off would mean they would be off for less than 15 minutes. They need extra electricity to start up and the bulb life can be shortened considerably by frequent starts.

2 • Use fluorescent or HID lamps wherever practical.

3 • Use three-way bulbs and install dimmers wherever practical.

4 • Keep all of your bulbs clean; the dirtier they are, the less light they give off.

5 • Use lower wattage bulbs for general lighting or as fillers.

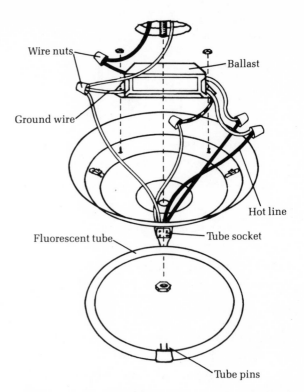

Wiring arrangement for an overhead fixture. Fluorescent bulbs come in either circular or tube shapes

6 • Install automatic switches in closets to make sure the light goes off every time the door is closed.

7 • Periodically check the lights in your attic, basement and/or any other out-of-the-way place to be sure they have not been left on inadvertently.

8 SOLAR ENERGY

Solar energy is on the way. There is already a proliferation of solar designs and systems, most of which are still too expensive for the average homeowner, and have too long a payback. The solar homes in existence also need a conventional heating system as a backup, which means there is the expense of double installation. But as the energy crunch goes on, solar energy will become more and more a way of life until, in the not very distant future, it may well become the primary method of heating our homes and public buildings. It is, without question, less expensive to operate once it has been installed, as well as cleaner and quieter than any of the mechanical systems. And there is virtually no maintenance involved.

In order to use the sun's rays for heating your home, you must have some way of collecting that heat, a method of storing it, and a system for distributing it throughout your home when you need it—which is usually when the sun is no longer shining. You can do this with either an *active* or a *passive* system. Furnaces, boilers, water heaters, heat pumps and air conditioners are active heating and cooling systems, because they require expensive energy-using equipment. An active solar heating and cooling system would also contain fans, pumps, storage or heat-exchange machinery and controls, all of which cost considerable amounts of money at this stage in the development of solar energy.

A passive solar energy system for heating and cooling relies on four things: the use of a natural energy (the sun); a minimum of mechanical parts or hardware; little or no energy; and it costs relatively little to install. An example of good passive solar energy design is the standard window. If a window is the correct size and type, is properly oriented to the sun and wind, and has an operable, insulated shutter, it can very often be used as efficiently as any existing active solar collector system.

It is the purpose of this chapter to merely suggest passive solar heating and cooling ideas, to present a few of the passive solar possibilities that you might consider as addenda to your present heating and cooling systems. If you have followed the suggestions in this book, you have already shored up your home about as much as you possibly can to conserve the energy you use. By adding a passive solar heating or cooling system, you can, without question, reduce your energy costs by substantial amounts. Here are some of the ideas and approaches

that are beginning to be used by solar energy pioneers.

MOVABLE INSULATION

Even if the windows in your house are triple glazed, they are virtually open holes for heat to enter and escape from the living areas. At night, or when the day is cloudy, the heat drain through your windows is astounding, so a great many people have taken to insulating their windows. Movable insulation can be placed over a window whenever the sun is not shining during the winter or if the day is too hot during the summer. You can design and build your own insulated shutters to fit over the windows in your home. Create a panel or curtain made of some kind

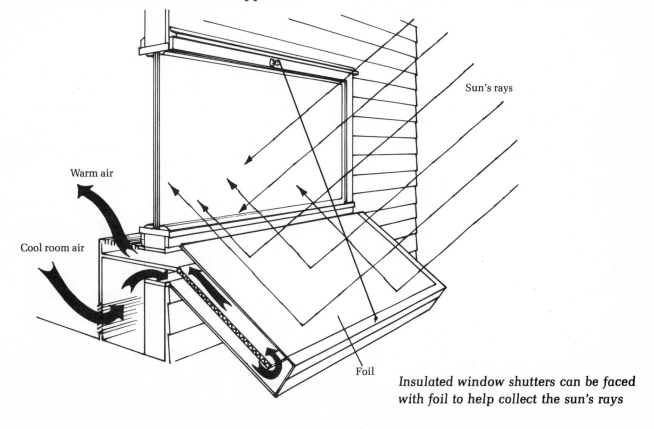

Insulated window shutters can be faced with foil to help collect the sun's rays

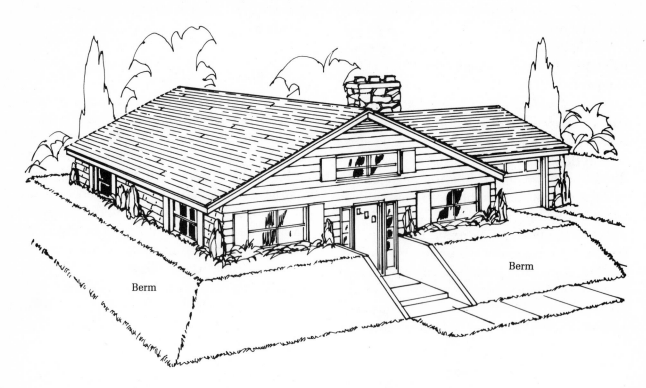

Berming can be so inclusive it practically buries the house

of insulating material, such as fiberglass, covered styrofoam, polystyrene, or just several layers of heavy cloth. When the shutter is in place over the window, it should fit snugly against the windowpane or flush against the frame with no cracks around its edges where heat can pass.

Insulating panels can be attached to the outside of the window or to the inside. They can be hinged or hung on hooks, held in place by magnets, folded open, slid back on tracks or operated by any other method you care to use. If you are covering the outside of a window, you might face the inside of the insulated panels with reflective foil and hinge the panel along its bottom so that to open it you must lower the top edge toward

the ground. When the panel is closed over the window, its insulation will keep heat in the house. When the panel is open, the reflective surface will help bounce extra solar heat into the house.

BERMS

Houses that are built to be heated by solar energy are often partially (or completely) buried in the ground, or *bermed*. Berming your house is to erect slanted banks of earth against its exterior walls. Not only will the earth protect the structure from winds, but a complete envelope of earth will guarantee a stable indoor temperature of 55° to 60°F every day of the year, because that is the temperature range of the earth below the frost line.

Almost any existing home can be bermed, but you will have to balance your desire for protection with your need for windows. You might consider burying the whole north side of your home while piling berms on the east and west sides only about waist high. But before you start berming, consult a structural engineer or architect who is familiar with berms. You have to be sure your walls are strong enough to bear the weight of all that earth. The walls also must be insulated on the outside as far down as 2' to 4' below the surface of the earth, or at least well below the frost line. You also have to install a special draining system in the berm so that the earth will not soak up rainwater

and hold it against the walls; the walls should be waterproofed as well as insulated. When construction of the berm is finished, landscape it, both for the sake of appearance and to hold the earth in place.

By making a modification of the berm you can incorporate a solar collector to store the sun's energy. Insulate and waterproof the outside of any wall as far up as the windows. Then construct a rock storage bin against the wall with warm air ducts under the window ledges leading into the house (they should have dampers). The berm is then piled against the storage bin and held in place by a glass solar collector. Cool air return ducts lead from the floor of the room under the berm to the bottom of the collector. The angled berm becomes a support for

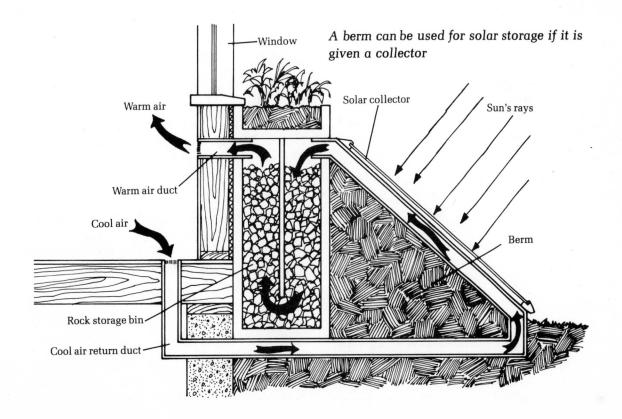

A berm can be used for solar storage if it is given a collector

Window

Warm air

Solar collector

Sun's rays

Warm air duct

Cool air

Berm

Rock storage bin

Cool air return duct

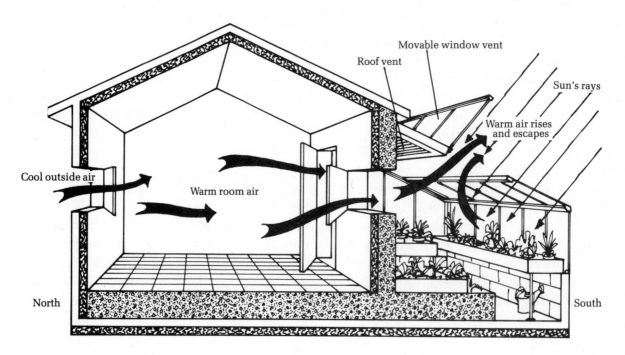

How a greenhouse can help cool a residence

the solar collector, which in turn sheds any rainwater that falls on it. The earth beneath the collector helps the rocks to store heat, which is ducted directly from the storage bin into the house. By closing the warm air ducts and shading the berm collector on hot days, cool outside air can be drawn into the house via the cool air return duct, allowing the berm to act as a cooling system during the summer.

SOLAR HOT WATER SYSTEMS

Presently, there are several solar hot water systems available to homeowners at a cost of approximately $2,500 to $3,000. In most cases, they consist of a collector, a water storage tank and a hot water heater (which runs on electricity) to function whenever

the sun fails to heat the water to the desired temperature or during peak demand periods. Some of the solar hot water systems can be united with your existing hot water heater. These units have a solar-heated tank of their own, which is used to feed the existing heater with heated or partially heated water. The solar hot water systems can be counted on to reduce the cost of heating your water by about 90% and because of their relatively low installation cost, they will have paid for themselves within about four years.

GREENHOUSES

If you attach a greenhouse to the side of your home, it will be as if you had added another wall beyond the outside of the house, and will serve to raise the temperature immediately beyond the exterior wall during the winter. If the greenhouse is filled with plants and can be properly shaded during the summer, it will also help to cool the interior of the house. Furthermore, all that glass aimed at the sun's rays automatically turns the greenhouse into a passive solar collector.

There is a wide range of solar greenhouse designs to choose from. Greenhouses can be attached to any exterior, but the preference is to put them along a wall facing south, where they can receive sunlight throughout most of the day. There are some other points to remember if you're planning to install a greenhouse with the idea of using it as a passive solar collector:

1 • In order to control the flow of heat and also keep the greenhouse from overheating, the wall between the greenhouse and your residence should be well insulated.

2 • Any surfaces that are directly exposed to the sun should be painted dark colors to absorb heat, while floors should be a light color to keep them from becoming too warm.

3 • Insulated shutters that can be closed over the greenhouse after the sun goes down will help retain the heat that collects during the day.

4 • There shouldn't be any trees or other structures outside the greenhouse that are closer than three times its height.

5 • You can improve the ability of the greenhouse to retain the heat it collects by building it on top of a rock storage bin. Just how much the storage bin will help

depends on the design of the greenhouse, the insulation used, and the design and location of your home.

With a greenhouse added to the outside of your home, you can open the doors, windows and vents between the house and the greenhouse during cold days to allow whatever heat has collected to flow into the main residence. Then shut the doors at night to prevent heat from flowing out of the house again. In the summer, you can use both high and low vents in the greenhouse to let hot air escape from the house and pull cooler air inside.

TROMBE WALLS

Developed by Dr. Felix Trombe and architect Jacques Michel, the Trombe wall is normally one entire side of a house, usually facing south. The wall must be made of concrete, stone or masonry, painted a dark color, and can be as much as 16″ thick. It may or may not have windows in it, but it must have vents placed regularly along the floor and ceiling of each room. Mounted several inches in front of the wall are one or two layers of glass.

During the winter, the vents are opened by day and closed at night. As the sun heats the air between the wall and its glass skin, the air rises and enters the house through the ceiling vents while cooler air is pulled into the wall's air space through the floor vents. At night, the top vents are closed so that the heat stored in the wall can radiate into the rooms.

During the summer, warm air in the house is drawn out through the top vents while

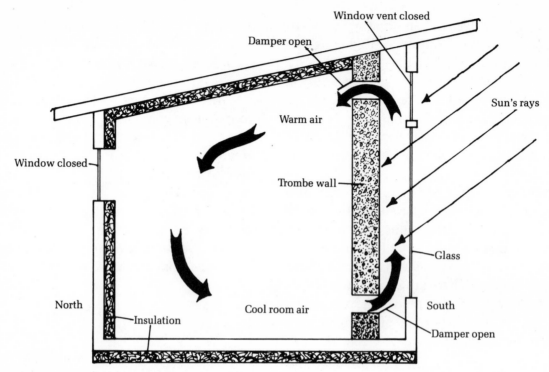

How a Trombe wall heats during a winter day

How a Trombe wall cools during the summer

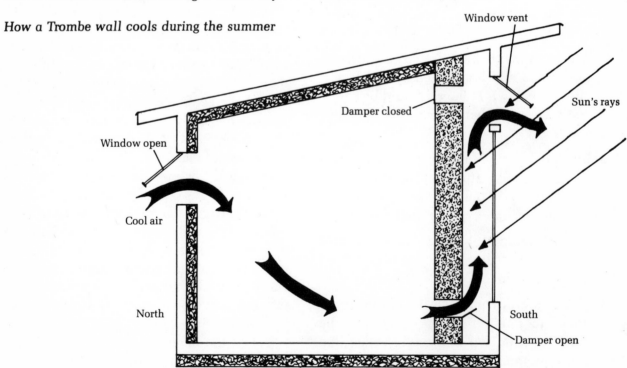

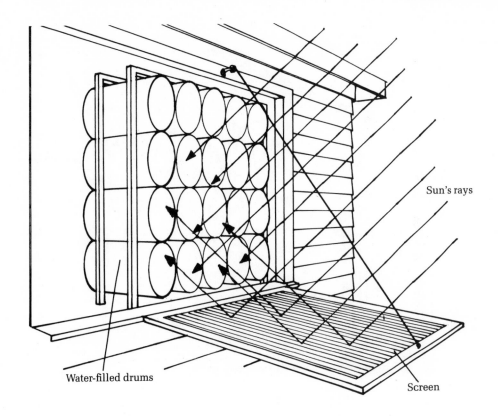

Sun's rays

Water-filled drums

Screen

A water wall made with steel drums

cooler outside air is pulled inside through the bottom vents. At night all of the vents are kept open to draw the warm air in the house out through the wall.

WATER WALLS

A variation of the Trombe wall uses water instead of concrete or masonry as its heat storage mass. The entire south side of a building is made with steel drums or vertical tanks made of highway culvert, steel, concrete or fiberglass, lined with polyethylene and filled with water. The reason for the polyethylene liners is to prevent leaks in the water containers caused by mineral deposits

in the water eating through them. Surfaces of the containers that face the sun are painted black or some other dark color. Glass is positioned a few inches in front of the water walls, and there should be an insulated shutter of some sort that can be opened and closed over the glass.

The particular advantage of a water wall over the Trombe wall comes from its use of water as a heat storage mass. Water is an infinitely more efficient storage media than concrete or masonry; it gathers heat faster and holds it for longer periods of time, as well as releases it more readily. Given a system of insulating panels and exterior shades of solar screens, you can shade the water

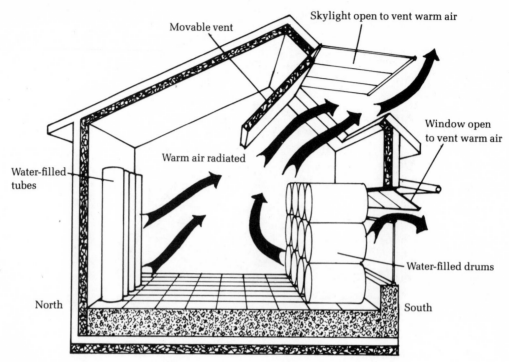

How a water wall cools during a summer night

How a water wall heats during a winter day

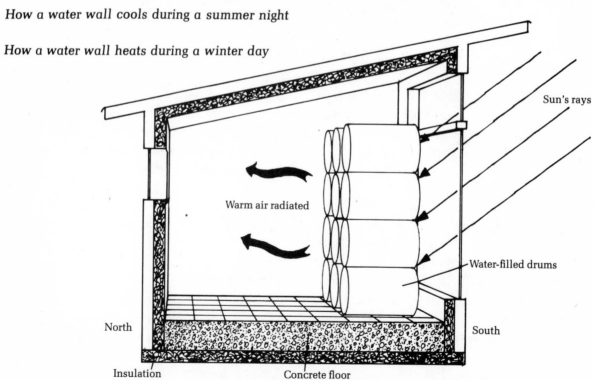

wall during hot days so that the water does not become heated and can function to cool the room behind it. It will keep the room temperature between 75° and 80°F when it is 115° outside. In the winter, the drums collect heat from the sunlight and then radiate it into the room at night when the outside wall is covered by its insulated screen.

ROOF PONDS

You have to have reinforced rafters for this, but a roof pond can do wonders for heating and cooling any home. Roof ponds are generally open troughs or polyvinyl (PVC) bags filled with water, which stores the solar heat. The ponds are placed under the roof, where they collect and store the sun's heat and then radiate it down to the rooms below.

There are some drawbacks to the roof ponds, however. The PVC bags will eventually disintegrate from the sun's ultraviolet rays, so they must be replaced occasionally, least they develop leaks. And unless you have a special ducting system to transfer heat to the lower floors of the house, the ponds will only serve those rooms immediately below them in the top story of the house.

Given the basic idea of roof ponds, there are several modifications for making them extremely efficient. You can install an insulating cover that rolls over the ponds at night to reduce heat loss through the roof. Or there can be an insulated lid over the ponds which has its inner surface covered with a reflective foil. The lid is cranked open during the day so that the foil bounces extra sunlight down to the ponds and in-

How a roof pond, using an insulated panel, heats during a winter day (left) and keeps heat in the house at night (right)

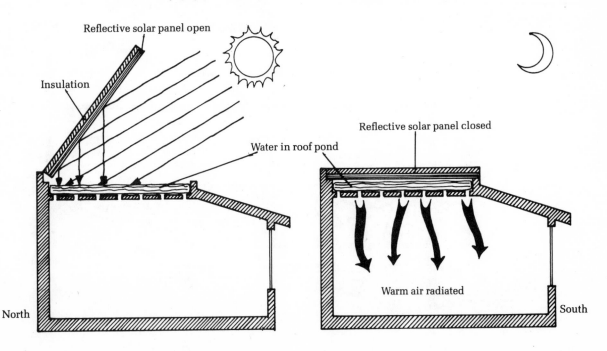

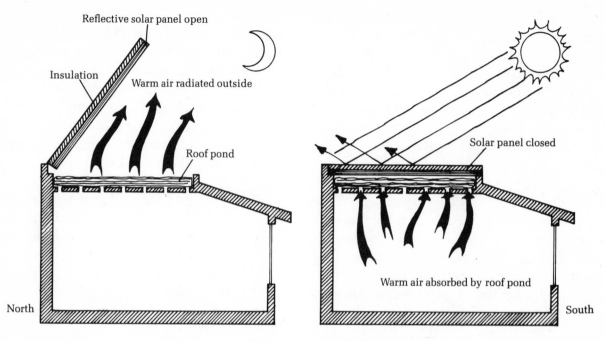

How a roof pond cools at night (left) and by day (right) during the summer

A thermosiphoning system. When the sun is not shining, the collector ducts can be closed off so that cool air in the house can return to the storage area to be warmed

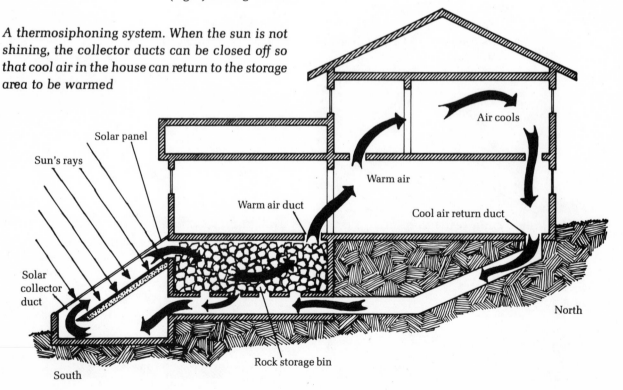

creases their heat intake. At night, the lid is closed to hold the heat in the water and help to reflect it downward into the house.

In climates that offer cool summer nights and low humidity, the insulated lid is opened after dark to release whatever heat is in the water. If the pond is made of open troughs, this cooling process is helped by evaporation of the water, but the system works almost as well when the water is stored in bags. During hot days, the lid is kept closed so that the water will absorb any heat rising from the house below.

THERMOSIPHONING

Thermosiphoning is the use of natural convection for distributing heat. In other words, it is the process of letting heat rise naturally from wherever it is collected to its storage point and then up through the house. Obviously, if the heat is going to rise up from its storage area, that area is not on your roof. The collection panels may be anywhere on the outside of the house, but the heat they gather is stored in rocks or water somewhere under, or at least downslope from the house. From the storage area, the heat is then ducted up through the rooms, rather than drawn up by mechanical means.

Thermosiphoning is ideal with a home that is situated on a hillside so that the collector can be stationed downhill from the storage bin, which is under at least part of the house. The collector can be as simple as a wooden frame supporting one or two layers of glass mounted in front of a black metal plate. It should be mounted at an angle to the sun and face south to catch the maximum rays from the low winter sun. As the sun heats the metal plate, air between the glass and the metal is also heated, causing it to rise to the top of the plate and into ducts leading to the storage bin under the house. The bin should be a massive area in the cellar filled with rocks, and have hot air ducts leading from it to whatever parts of the house are to be heated. Warm air simply rises through the ducts, into the house. There also must be return air ducts that will bring cooler house air back down to the collector where the sun can heat it. Ducts must have dampers so that when the sun is not shining, the collector can be closed off from the house and storage bin. At those times, the cooler air in the house will be able to return only as far as the storage bin, where it is warmed by the rocks, and then rises up through the house again. During the summer, the top of the collector should be vented to the outside and closed off from the storage bin to prevent the rocks from heating.

THE SOLAR FUTURE

There are already numerous passive and active solar systems in existence and more being invented every day. Some of these systems can be inexpensively installed by a homeowner. Others require the purchase of costly hardware. But sooner or later, most American homes will become involved with some form of solar collection and storage. Exactly when each family enters the world of solar heating will vary, of course, and so will the degree of participation. People who are energy-conscious and presently in the process of building new homes are likely to be incorporating some form of solar assistance right now. Others are retrofitting their present residences to passively heat part of their homes or to heat their water. The majority of American homeowners are investigating the possibilities and watching the developments in solar energy. Already the federal government, as well as many state and local governments, are allowing tax deductions to defray the cost of installing a solar energy system. It probably won't be very long before you too become involved in the use of solar heat as a way of reducing the rising costs of heating and cooling your home. It will not be very long before solar energy becomes a major avenue for meeting most of our energy needs.

9 seasonal ENERGY-SAVING checkups

I f you were to shut off all of your utilities, lock up your house, and go away for one year, you would, upon your return, be amazed at how much the building had deteriorated just standing there unused for twelve months. Wood, which is the principal material in most resident constructions, is constantly changing its dimensions. It swells in damp or humid weather, shrinks in dry, cold air, and can change its shape as often as four times an hour if it happens to be in a steam-heated room on a cold day. Each time the wood in your home swells or shrinks, it changes the size of the joints between it and other materials—around the perimeter of an aluminum window frame, for example. Given a year of inattention a house can develop cracks in its walls and ceilings, and drafts around all of its windows and doors. The roof can begin to leak, cement can crack, pipes with water left in them can burst. The list of maladies is almost endless, and each new season of the

year can aggravate or instigate still more problems. A house—any house—is a living, breathing entity, and as such it must be cared for on a regular basis. This means it should be thoroughly inspected at least twice a year and any imperfections you find should immediately be corrected before they can develop into major problems. The two best times of the year to make your semiannual house check are in the spring and the fall, when you can open and close the windows without losing inordinate amounts of heat or cool. Not all of the areas you inspect pertain directly to saving heating and cooling costs, but it is important to keep them repaired so that damage does not affect the energy efficiency of your house. A leaking roof, for example, does not directly change the ability of your home to retain heat. But at the point where water works its way into the attic insulation and diminishes its insulating properties, that hole in the roof begins to cost you time and money.

SPRING CHECKLIST

ROOF

1•Examine the roofing carefully for loose or damaged shingles. Apply roofing cement under loose shingles and renail them. Replace damaged shingles.

2•Check the flashing around the chimney, vent pipes, dormers, hips and valleys. Repair any damaged areas with roofing cement, or replace the flashing.

3•Clean the gutters and downspouts of all sludge, debris, etc.

4•Look at the gutters and downspouts for leaking joints or holes. Repair or replace all damaged areas.

5•Check the downspouts for loose connections and clogging. Clean the leaf strainers if there are any; tighten any loose seams and caulk any areas that leak.

6•Test all of the masonry in the chimney for loose pieces and defective joints. Repair damaged areas with mortar.

7•Clean all roof vents. If there is a fan on the vent, be sure it is able to move freely.

8•Clean the inside of the chimney.

HOUSE EXTERIOR

1•Go over all the siding on your house and note any problems with ventilation or moisture. Repair or replace damaged siding; paint the house if it is necessary.

2•Fill all cracks around the door and window trim with caulk.

3•Examine all masonry work for water damage and cracks. Fill all cracks and holes with the appropriate patching material.

4•Inspect all door, window and porch screens for tears or wear. Repair or replace the screening as needed.

5•Remove all storms, mark their location and store them carefully.

6•Examine all screens and their frames for damage. Repair any cracked or broken frames.

7•Clean all screens and put them up.

8•Check storm and screen doors to be sure they open and close properly. Tighten loose hinges; repair damaged areas.

9•Inspect all door closers and adjust them to close their doors properly. Replace any closers that cannot be repaired.

10•Take awnings out of storage and check them for damage. Repair any tears or holes and install the awnings.

11•Inspect the outside frames of windows closely. Repair loose putty around the panes, fill cracks in the frames and paint.

12•Remove the winter covers on air conditioners and service the units. Clean their evaporator coils and unclog drain lines.

13•Plant any shrubs or trees to be used as shade for your windows.

14•Clean your central air conditioner if you have one. Be sure the base is not cracked or damaged and that no grass is growing in front of the coils.

15 • Replace all defective weather stripping around doors and windows.

16 • Be sure that the grading around the foundation slopes *away* from the house. If it slopes toward the house, build it up so that there is a proper slope to carry water away from the basement walls.

17 • Inspect the garage door and lubricate it if necessary. Replace defective weather stripping, glass and hardware.

BASEMENT

1 • Examine the joint between the floor and walls for any signs of dampness. Patch any cracks or holes.

2 • Examine the foundation walls for cracks or holes and dampness. Repair as needed.

3 • Check along the sill for air leaks. If there is no insulation between the top of the foundation walls and the first floor, install batts with the proper R-value for your area.

4 • Check for air leaks around all basement windows. Install weather stripping.

5 • Inspect the basement windows for damaged frames or broken glass. Repair all damaged areas. If you have casement windows, lubricate their crank mechanisms.

6 • Look at the insulation between the floor joists. If there are sags anywhere, add wire or wooden bracing to hold it up between the joists.

7 • If you have no insulation under the basement floor, you should at least have it on

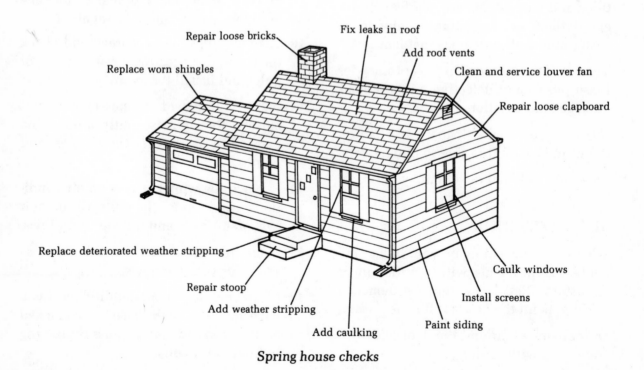

Spring house checks

the basement walls. Insulation often goes on sale in the spring; now is a good time to buy and install it. Keep your purchase bills—insulation is now a federal tax deduction.

8•Check all plumbing joints for leakage. Repair if necessary.

9•Wrap the hot water supply line in insulation if it is not already wrapped. If the hot water line is wrapped, inspect the insulation for tears or holes and repair any damage.

10•Run your hands along the ducts in a forced air heating system while the furnace is running to check for air leaks. Cover leaks with duct tape.

11•If the duct system in a forced air arrangement is to be used for central air conditioning, it should be covered with insulation. If it is already insulated, inspect the insulation for any damage and repair as necessary. Wrap the ducts in insulation if none exists.

12•Clean and lubricate the sump pump.

CRAWL SPACE

1•Check the foundation wall and pylons for damage and repair if necessary.

2•Be sure that all insulation and vapor barriers are intact. Repair or replace if necessary.

3•Be sure that any form of drainage under the crawl space is unclogged and functioning.

SLAB

1•Examine all masonry for cracks or holes and repair.

2•If possible, check all insulation for water damage. If you find any, trace the leak and make the appropriate repairs.

FURNACE

1•Clean all filters and replace if necessary.

2•Clean the humidifier if there is one.

3•Clean the blower.

4•Clean and lubricate the blower motor.

5•Drain the boiler and radiator pipes if your heating system uses hot water or steam. Replace with clean water.

6•Check the T&P (temperature/pressure) valve; replace if defective.

7•Check the stack control and clean its heat sensor.

8•Turn off all thermostats.

9•Clean and lubricate the pump and its motor.

10•Make an appointment with qualified service personnel to clean and check your furnace and chimney.

HOT WATER HEATERS

1•If your hot water heater is gas- or oil-fired, clean the flue.

2•Examine the burner on either type and clean it if necessary.

3•Drain water from the bottom of the tank until it turns clear.

4•Test the T&P valve; replace if it is defective.

5•Withdraw the anode rod and inspect it. If it is nearly all corroded, replace the rod.

6•If you have a timer controlling the unit, reset it to coincide with the longer summer days.

APPLIANCES

1•Clean and lubricate all fans, including the attic fan, room fans and the one in your stove hood. Inspect each unit for worn or broken parts and repair or replace.

2•Inspect the seals on your refrigerator-freezer doors and repair or replace them.

3•Clean the coils on the back of the refrigerator.

4•Clean all room air conditioners and change their filters.

5•Be sure that the seal around each air conditioner is airtight.

6•Clean and service your dehumidifier according to the manufacturer's instructions.

7•Be sure the exhaust on your clothes dryer is clean and venting properly.

GENERAL

1•Repair all leaking faucets by replacing their seat washers.

2•Clean fireplaces thoroughly. If you do not have central air conditioning, a good way of improving ventilation in the house is to leave the damper open.

3•Check all radiators for leakage around their service pipes and repair or replace any connections that are faulty.

4•Bleed all steam or hot water radiators.

5•Remove and clean the air valves on steam radiators by soaking them overnight in vinegar.

6•Remove all registers in a forced air system and clean the ducts behind them, particularly if the ducts are used for central air conditioning.

7•Check all blinds, shades and shutters to be certain they are in good working order. Clean them thoroughly.

ATTIC

1•Go through your attic very carefully to find any signs of moisture. If the attic is unfinished, look carefully at the insulation in the floor to be sure it is dry. Trace any moisture back to its source and fill the leak.

2•Clean all vents and be sure they are bug-free. Oil and clean powered vents according to their manufacturer's instructions.

3•Clean the attic to remove as much dust and dirt as possible.

FALL CHECKLIST

Some of the energy-saving maintenance chores for the fall are repeats of the spring, primarily because such things as caulking can deteriorate in less than a six-month span, or may not have been noticeable in the spring but have begun to develop into problems after a summer of hot, humid weather. Don't cheat yourself out of savings on your energy bills by skipping parts of either the spring or fall checkup.

ROOF

1•Check for loose or damaged shingles and make the appropriate repairs.

2•Examine the flashing for leakage and coat with roofing cement.

3•Clean the gutters and downspouts of all leaves and debris.

4•Examine the gutters and downspouts for leaks and repair.

5•Inspect the eaves to be certain the drip edge is secure and the eave shingles are in good repair. They may have to bear the weight of a great deal of snow in the coming months.

HOUSE EXTERIOR

1•Examine the entire outside of your house for leaks, cracks, any sign of moisture or dampness. Repair as needed.

2•Examine the caulking around all windows and doors and repair as needed.

3•Check weather stripping around windows and doors. Replace any that is less than perfect.

4•Put the winter covers on your exterior air conditioners; cover all room air conditioners that are to remain in place, or remove and store them.

5•Remove all screen doors and windows and store them.

6•Inspect all storm doors and windows for broken glass and weathered frames. Repair as necessary.

7•Install your storm windows and doors. Add caulking or weather stripping to make them as airtight as possible.

8•Examine the door sweeps on all outside doors. Replace any that are not in perfect condition.

9•Check the door closers on all outside doors. Repair or replace faulty closers.

10•Put away awnings.

11•Look at the outside faucets to be sure the caulking around their bases is solid and airtight.

12•Drain the water out of all outside pipes.

BASEMENT

1•Check the walls and floor for any signs of water damage. Locate all leaks and repair damaged areas.

2•Examine the insulation around the sill and between the floor joists or on the walls. Be sure it is in its proper place and not damaged.

3•Check around all basement windows for air leaks. Install weather stripping or caulk to eliminate leakage.

4•If you have not done it before this, wrap all exposed hot water pipes with pipe insulation. If the pipes are already insulated, make sure the insulation is in good repair.

5•Check all plumbing joints for leakage and make the necessary repairs.

6 • Test the hot air ducts for leakage and re-pair with duct tape.

7 • Examine all exposed hot water or steam pipes for leakage and repair.

8 • Clean and lubricate the sump pump.

CRAWL SPACE

1 • Examine the foundation walls and pylons for damage and repair.

2 • Look at the insulation and vapor barrier to be sure they are not damaged and are in the proper places. Replace if necessary.

3 • Check the drainage system to be sure it is clean and functions freely.

SLAB

1 • Examine all masonry for cracks or holes and repair.

2 • Check as much of the insulation as you can for water damage. Locate the source of the damage and repair it. If possible, replace the damaged insulation.

FURNACE

1 • Make sure the furnace has been cleaned and serviced by a qualified serviceman.

2 • Check all filters and clean them if nec-essary.

3 • Be sure the humidifier is clean and in good working order.

4 • Clean the blower, particularly if it was used during the summer.

5 • Clean and lubricate the blower motor.

6 • Check the water content in hot water and steam heating systems. It should be clean. If it is not, drain the boiler and pipes and replenish the system with fresh water.

7 • Test the T&P valve. Replace if faulty.

8 • Adjust the thermostats to their correct winter settings.

9 • Clean and lubricate the pump on hot water systems.

10 • Be sure you have enough reserve fuel.

HOT WATER HEATERS

1 • Be sure the flue is clean if your heater is gas- or oil-fired.

2 • Examine the gas or oil burner to be sure it is clean and working properly. If not, clean it or have it serviced.

3 • Drain water out of the tank until it turns clear.

4 • Test the T&P valve. Replace if faulty.

5 • Check the anode rod. Replace if exces-sively corroded.

6 • If the heater is controlled by a timer, reset it for winter hours.

APPLIANCES

1 • Clean and put away all room fans. Clean the fan in your stove hood and repair or replace any broken or worn parts.

2 • Clean the coils on the back of the refrig-erator.

3 • Clean all room air conditioners and cover

them for the winter, or store them.

4•Put away your dehumidifiers.

5•Clean and service all room humidifiers.

6•Clean all lights, particularly the overhead fixtures, so they can provide more light at less wattage.

GENERAL

1•Repair leaking faucets by replacing their seat washers.

2•Check the dampers on fireplaces to be sure they close tightly and admit no drafts.

3•If you do not have any heat dispersing equipment for your fireplaces, buy some. The cost of the purchase will be returned to you within a season of lowered fuel bills.

4•Check radiators for leakage and repair.

5•Be sure all radiator valves are functioning properly. Clean or replace faulty valves.

6•Clean behind all forced air registers, especially if they were used for central air conditioning during the summer.

7•Check blinds, shutters, shades and curtains to be sure they are working properly and are clean.

ATTIC

1•Check your attic thoroughly for any signs of moisture or dampness. Feel the insulation and peer up at the roof sheathing. If the insulation is wet, replace it; if you see daylight coming through the roof, repair the crack.

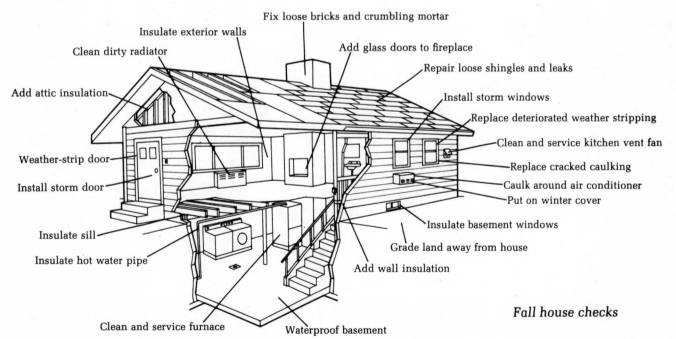

Fix loose bricks and crumbling mortar

Insulate exterior walls

Clean dirty radiator

Add glass doors to fireplace

Repair loose shingles and leaks

Add attic insulation

Install storm windows

Replace deteriorated weather stripping

Clean and service kitchen vent fan

Weather-strip door

Replace cracked caulking

Install storm door

Caulk around air conditioner

Put on winter cover

Insulate basement windows

Insulate sill

Grade land away from house

Insulate hot water pipe

Add wall insulation

Fall house checks

Clean and service furnace

Waterproof basement

index